Extremes in Random Fields

Extremes in Random Fields

A Theory and its Applications

Benjamin Yakir

Department of Statistics
The Hebrew University of Jerusalem, Israel

WILEY

HIGHER EDUCATION PRESS

Library of Congress Cataloging-in-Publication Data
Yakir, Benjamin, author.
  Extremes in random fields : a theory and its applications/Benjamin Yakir.
     pages cm
  Includes bibliographical references and index.
  ISBN 978-1-118-62020-5 (hardback)
  1. Random fields. I. Title.
  QA274.45.Y35 2013
  519.2′3–dc23

                                                                        2013018539

A catalogue record for this book is available from the British Library.

ISBN: 978-1-118-62020-5

Set in 10/12pt Times by Laserwords Private Limited, Chennai, India
Printed and bound in Singapore by Markono Print Media Pte Ltd

1 2013

To David

Contents

Preface

This text started as class notes for a course that I gave in the Mathematical Sciences Center (MSC) in Tsinghua University, Beijing, that got overblown and became a book. I was enjoying a sabbatical leave in the Department of Statistics and Applied Probability (DSAP) of the National University of Singapore when I was given an offer to teach a summer course in China. Of course I accepted. How could I resist the opportunity to fulfil a childhood dream of visiting China?

After accepting the proposal I had to decide what to teach. I decided to fulfil yet another dream, the dream of summarizing and unifying a subject I was writing about all my career, even before I knew what the subject was. The subject is the distribution of extremes in random fields and the analysis of statistical problems that can be formulated in relation to such extremes. Immediately after obtaining my PhD, and as a continuation of my PhD thesis, I was interested in the investigation of the average run length of the Shiryaev–Roberts change-point detection rule. Therefore, I found it natural to try to address a challenge that was presented to me by David Siegmund during a barbecue meal that he prepared for me in his yard. The challenge was to develop a simpler method for analyzing this average run length. In an attempt to attack this problem I began experimenting with the likelihood ratio identity, one of David's favorite techniques, and followed the road that eventually led me to writing this book.

The original problem was the investigation of average run length in a sequential change-point detection problem.[1] However, the basic technique that was developed turned out to be useful for the investigation of a relatively wide array of different statistical problems that involve the distribution of maxima.[2] Among other things, David and I used the method in order to investigate the significance level of sequence alignment, for the computation of the false detection rate in

---

[1] Yakir B., Pollak M. A new representation for the renewal-theoretic constant appearing in asymptotic approximations of large deviations. Ann. Appl. Probab. **8**, 749-774 (1998).

[2] Grossman S., Yakir B. Large deviations for global maxima of independent superadditive processes with negative drift and an application to optimal sequence alignment. Bernoulli **5**, 829–845 (2004).
Seigmund D.O., Yakir B. Approximate p-values for local sequence alignments. Ann. Statist. **28**, 657–680 (2000).
Seigmund D.O., Yakir B. Statistical analysis of direct identity by descent mapping. Ann. Hum. Genet. **67**, 464–470 (2003).
Seigmund D.O., Yakir B. Correction note: Approximate p-values for local sequence alignments. Ann. Statist. **31**, 1027–1031 (2003).

scanning statistic, for producing more efficient ways of simulation, etc. Each application required this modification or that trick in order to apply the basic principle. However, after 20 years of repeating the same argument even I was able to identify the pattern. The thrust of this book is a description of the pattern and the demonstration of its usefulness in the analysis of nontrivial statistical problems.

The basic argument relies on a likelihood ratio identity that uses a sum of likelihood ratios. This identity translates the original problem that involves the approximation of a vanishingly small probability to a problem that calls for the summation of approximations of expectations. The expectations are with respect to alternative distributions in which the event in question is much more likely to occur. Moreover, by carefully selecting the alternative distributions one may separate the leading term in the probability from the expectations that form the sum, enabling the investigation to concentrate on finer effects.

The method is useful since it does not rely on the ordering of the parameter set and it does not require the normal distribution. In many applications, some of them are presented in the book, a natural formulation of the model calls for the use of collections of random variables that are parameterized not by subsets of the real line. Frequently, the normal assumption may fit the limit in a central limit formulation but may not fit as a description of the extreme tail. In all such cases an alternative to the methods that are usually advocated in the literature are required. The method we present is such an alternative which we felt others may benefit from by knowing about.

This is why we wrote the book. But who is the target audience? This is a tough call. Even if I may state otherwise, the book requires a relatively advanced knowledge in probability as background, perhaps at the level of Durrett's book.[3] Prior knowledge in statistics is an advantage. Indeed, there is an appendix that lists theorems and results and can be used as reference for the statements that are made in the book. Still, I guess that this book is a not an easy read even for experts, and much less so for students.

With this warning in mind, I hope that the effort that is required in reading the book will be rewarding. Definitely, for an expert who wants to add yet another method to his toolbox but also for a student who wants to become an expert.

Seigmund D.O., Yakir B. Significance level in interval mapping. In Development of Modern Statistics and Related Topics, Series in Biostatistics, Volume 1. World Scientific Publishing, River Edge, NV, 10–19 (2003).

Shi J., Siegmund D.O., Yakir B. Importance sampling for estimating p-values in linkage analysis. JASA **102**, 929–937 (2007).

Yakir B. On the average run length to false alarm in surveillance problems which possess an invariance structure. Ann. Statist. **26**, 1198–1214 (1998).

Yakir B. Approximation of the p-value in a multipoint linkage analysis using grandparent grandchild pairs and partially informative markers. Nonlinear Anal. **47**, 1973–1984 (2001).

Yakir B. Discussion on "Is average run length to false alarm always an informative criterion?" by Yajun Mei. Sequential Analysis **27**, 406–410 (2008).

[3] Durrett R. Probability: Theory and Examples (2nd Edition). Duxbury Press, Belmont, CA (1995).

For such students, the book can be used as a basis for an advanced seminar. Reading chapters of the book can be used as a primer for a student who is then required to analyze a new problem that was not digested for him/her in the book. This is how I intend to use this book with my students.

The teacher can start such a course by discussing Chapters 1–4 that give the basic background and demonstrate the technique. Chapter 5 is more technical and can be skipped, unless the main interest is in the mathematical details. From the second part of the book it is probably recommended to go over Chapter 6, which is of an intermediate level of difficulty, and then read some or all of Chapters 7–10 depending on the interests of the teacher and the students and on the time constraints.

Acknowledgments

My first acknowledgments are to environments, and especially the people who enabled these environments. The first half of the book was written mainly in DSAP. I know of no better place to do this type of scientific work. I will always be grateful. The second place is MSC. Without them I do not know when, if at all, this book would have been written. Next, I would like to recognize the financial support that I got from the Israel Science Foundation (Grant No. 325/09) and from the US–Israel Binational Science Foundation (Grant No. 2006101). This support was instrumental for the development of the original work that led to the applications that are presented in the second part of the book.

Some of the people that gave me a helping hand I would like to mention by name. Unfortunately, I cannot give the names of the anonymous reviewers who made very useful suggestions on the first draft of the book and helped me improve it. But I can give the name of the editor from Higher Education Press, Liping Wang. Thanks to her this is a book and not just class notes. Also I would like to thank Yuval Nardi, Moshe Pollak, Ton Dieker and Nancy Zhang who coauthored with David Siegmund and myself some of the works that are related directly to the content of the book.

And finally there is David Siegmund. The work presented in this book is basically our joint work. The only reason that we do not share authorship is the fact that I wanted to dedicate this book to him as my modest contribution to the celebration of his career and his accomplishments and as an appreciation for what he gave me. It is not appropriate for a book to be dedicated to one of its authors. So here it is: this is for you, David.

February 2013   Benjamin Yakir, Jerusalem, Israel

Part I
THEORY

Part 1

THEORY

1

Introduction

1.1 Distribution of extremes in random fields

The aim of this book is to present a method for analyzing the tail distribution of extreme values in random fields. A random field can be considered as a collection of random variables $\{X_t : t \in T\}$, indexed by a set of parameters T. The index set T may be quite complex. However, in the applications that we will analyze in this book it will typically turn out that T is a 'nice' subset of $\mathbb{R}^d$, the d-dimensional space of real numbers.

In some statistical applications one is interested in probabilities such as:

$$\mathrm{P}\left(\sup_{t \in T} X_t \geq x\right),$$

the probability that the maximum of the random field exceeds a threshold x, for large values of x. There are only a few special cases in which the problem of computing such probabilities has an exact solution. In all other cases one is forced to use numerical methods, such as simulations, or to apply asymptotic approximations in order to evaluate the probability. This book concentrates on the application of the proposed method for producing asymptotic analytical expansions of the probability. Nonetheless, some elements in the method may, and have been, applied in order to simulate numerical evaluations more efficiently. An application that illustrates the usefulness of the method in the context of simulations is presented in the second part of the book.

As a motivating example consider scanning statistics. Scanning statistics are used in order to detect rare signals in an environment contaminated by random noise. For example, let us assume measurements that are taken in a one-dimensional environment. Each measurement is associated with a point in the environment and the points are equally spaced. For the most part, the expected values of the observations are fixed at some baseline level throughout

Extremes in Random Fields: A Theory and its Applications, First Edition. Benjamin Yakir.
© 2013 by Higher Education Press. All rights reserved. Published 2013 by John Wiley & Sons, Ltd.

the environment. However, at some unknown locations the expected value is different from the baseline. Such a shift of the expectation extends over an interval of unknown length. An interval of shifted expectations is the signal we seek to identify. Such a signal is parameterized by the location of the interval, by the length of the interval, and perhaps also by the magnitude of the shift.

The expectations of the observations correspond to signals (or lack thereof). A complication in fulfilling the task at hand is the fact that the observations are subject also to random noise, which may be parameterized by the variance of the observations. Frequently, this random noise is taken to be normally distributed and independent among observations. In such a case, the expectation structure and the variance specifies completely the distribution of the observations.

Say that our goal is to decide whether or not there is any signal in the environment. A reasonable approach, which has statistical merits to it, is to associate with each potential signal a statistic that summarizes the information in the data regarding that signal. For example, if signals are all of the form of an interval with a fixed level of the expectation above the baseline then an appropriate statistic is the standardized sample average of the observations that belong to the interval, with standardization conducted with respect to the baseline expectation and variance. The presence of a signal in the environment is announced if there exists a statistic with a value above a previously determined threshold. False detection occurs when all observations share the same background expectation level but, due to random fluctuations, the threshold is crossed. The preliminary task of the statistician, in order to limit the probability of false alarms, is to determine the value of the threshold.

In the current example the parameter set T corresponds to the collection of all potential signal intervals. An element in T, an interval, is composed of a pair of numbers – the central location of the interval on the line and the length of the interval. Hence, T is a subset of the two-dimensional plain of real numbers. The statistic X_t that is associated with $t \in T$ is the standardized sample average based on the observations that belong to the interval t. Note that although the original observations were assumed to be independent, the collection of statistics $\{X_t : t \in T\}$ are not since two overlapping intervals share some of the observations. On the other hand, if the noise is normal then the distribution of the statistics is also normal. A graphical illustration of the situation is given in Figure 2.5 in Section 2.3.[1] Denote the threshold by x. It follows that the probability of false detection is of the form:

$$P\left(\sup_{t \in T} X_t \geq x\right),$$

which is the form we declared to be of interest for us. Approximations that relate the probability to the value of the threshold x enables one to select the value of the threshold to meet tolerated levels of the probability.

---

[1] In Figure 2.5, our t is denoted there by $\theta = (t, h)$, with t the central location and h the length of the interval. The statistic that we denote here by X_t is denoted there by Z_θ.

The text is devoted to the task of analyzing the tail probability of extremes. We ignore completely more fundamental issues of sample-path properties of the random field and questions of measurability of the random variable $\sup_{t \in T} X_t$. The parameter set T in many of our applications is either finite or countable. In such a case measurability of the supremum follows readily. In other cases one may rely on the separability of the random field in order to establish the measurability requirement. As for us in this book, we just ignore the issue.

The analysis of the probability that the maximal value of the field exceeds the threshold x in the situation where T is such that this probability is vanishingly small will occupy a central part of the discussion. This type of analysis, the analysis of vanishingly small probabilities, is frequently referred to as *large deviations*. Typically, a statement of a theorem in large deviations establishes the exponential rate by which the probability converges to zero. This first-order approximation of the probability will not be sufficient for our needs. The aim of our analysis will be to produce refined approximations, approximations that include polynomial terms and associated constants. These refined expansions open the door for the production of approximations to probability of events that involve maximization of a random field in settings where probabilities do not converge to zero or the computation of other functionals that are associated with such events.

For example, in the case of a scanning statistic with normal noise we have that the marginal distribution of each element in the random field is standard normal. Consequently,

$$P(X_t \geq x) = 1 - \Phi(x) \approx \frac{1}{x\sqrt{2\pi}} e^{-\frac{1}{2}x^2} ,$$

where Φ is the cumulative distribution function of the standard normal distribution and the approximation is valid for large values of x. It can be shown that

$$\lim_{x \to \infty} x^{-2} \log P \left(\sup_{t \in T} X_t \geq x \right) = -\frac{1}{2} ,$$

which is the content of the large deviation statement in this case. However, in Chapters 2 and 3 we prove the more detailed approximation:

$$P \left(\max_{t \in T} X_t \geq x \right) \sim x^3 e^{-\frac{1}{2}x^2} (2\pi)^{-\frac{1}{2}} \cdot (0.5)^2 (t_1 - t_0)(1/h_0 - 1/h_1) ,$$

when $T = [t_0, t_1] \times [h_0, h_1] \subset \mathbb{R}^2$. This more detailed description involves both the large deviation rate (appearing in the approximation in the form of the element $e^{-\frac{1}{2}}$) but also polynomial terms and constants.

In this book we will selfishly concentrate on a specific approach for dealing with the problem at hand and thus portray the false image that the method that we present is the best method, not to say the only method, for producing asymptotic approximation of the probability that a random field obtains an extremely high value. The special situation where the index T is a subset of the real line, in which case the random field is actually a random process, has a long history and many tools for solving. Some of the alternative methods of solution in this case will be presented briefly in the next chapter. Another notable special case with a very elegant theory is the situation where the random field is Gaussian with a smooth covariance structure. A more general tool may be applied in the Gaussian setting that involves a continuous parameter set T for cases where the covariance function does not have derivatives. There are far fewer tools available in order to deal with the even more general case where the random field is not Gaussian, and to handle in this non-Gaussian setting cases where the index set is discrete or the sample paths are not smooth.

The toolbox of the probabilist is not completely empty when faced with these more general problems. However, the optional methods are limited in number and none is very elegant. Admittedly, one may question the elegancy of the method that we will advocate. Still, the method seems to work in many different settings and thus may claim the title of generality.

The method is defined through a series of steps in which the large deviation part of the probability is accounted for first, followed by refinements that result from the identification of the contributions that are due to global and to more local fluctuations. These recommended steps may help to organize the investigation of the probabilistic characteristics of the problem at hand and assist in the evaluation of the relative contribution of the various sources of variability.

The demonstration of methods for approximating the extreme tail of the distribution of the maxima of a random field is initiated in the next chapter. The current introductory chapter is devoted to mental preparation. In Section 1.2 we provide a bird's view of the proposed method. In that section we outline the role and characteristics of the different steps. All the details are left out and are given in subsequent chapters.

Section 1.3 presents the type of random fields in which one may hope to apply the method as it is presented in the current text. Essentially, we are motivated by the analysis of Gaussian random fields, yet the the method is marketed as a tool that works for non-Gaussian fields as well. Nonetheless, the approximation is based to some extent on the application of the central limit theorem. Consequently, the type of fields that we target are those that obey the central limit theorem in an appropriate sense. We discuss such fields in Section 1.3.

In Section 1.4 we give a list of applications. Random fields associated with two relatively simple applications will serve throughout the first part of the book as a demonstration of classical methods for approximating the distribution of extremes, as well as the demonstration of our method. Other, more complicated examples will be discussed in the second part of the book, the part devoted to applications. For these examples we use the method described in Section 1.2.

1.2   Outline of the method

The method we propose is motivated by statistical considerations. We are inspired to think of the parameter t of the field as specifying a statistical model and consider X_t as a statistic that summarizes the information regarding the parameter. In many of the applications that we will consider this indeed is the context in which the field emerges. In other applications, when this is not the case, we may still consider that point of view as a motivation and a guiding principle.

More specifically, we propose to consider the problem of finding the tail distribution of a random field in the context of statistical hypothesis testing. In statistical hypothesis testing competing models for the distribution of the observations are grouped into two sub-collections. One sub-collection is called the null hypothesis. It reflects the absence of a scientifically significant signal in the data. The other sub-collection is composed of alternative distributions which are a reflection of the presence of such a signal. A statistical test is constructed with the role of determining which of the two hypotheses is more consistent with empirical observations.

Inspired by that point of view one may regard the random variable $\sup_{t \in T} X_t$ as the test statistic, with the test itself rejecting the null hypothesis in favor of the alternative if this test statistic is above a threshold x. The null hypothesis itself in this context is composed of a single distribution: the actual distribution of the data. Consequently, rejecting the null hypothesis is an error. The probability that we seek is the probability of making such an error, which in statistical vocabulary is called the significance level of the test.

The alternative collection of models is associated with the set T. Each $t \in T$ specifies a model P_t of the distribution of the data. This distribution may be the actual distribution that was considered, if the problem emerged in the considered statistical context, or it may involve some artificially constructed model that fits our needs. At the heart of the method is the proposal to translate the original problem of computing the probability of being above the threshold under the null hypothesis to a problem of computing expectations under alternative models. The vehicle that carries out this translation is the likelihood ratio identity.

The likelihood ratio identity employs likelihood ratios. A likelihood is the probability of the observed data under a given probabilistic model. If the distribution of the data is continuous then the likelihood refers to the probability's density. A likelihood ratio is the ratio between two likelihoods. Here we consider likelihood ratios in which the denominator is the likelihood of the data under the current distribution (the null distribution) and the numerator is the likelihood of the data under the alternative distribution P_t. We denote this likelihood ratio by $\exp\{\ell_t\}$, with ℓ_t being the log-likelihood ratio. We relate each X_t to ℓ_t, whether or not X_t emerged originally as a log-likelihood ratio, and rephrase the original problem of crossing the threshold by elements from the collection $\{X_t : t \in T\}$ using instead elements from the collection $\{\ell_t : t \in T\}$, possibly with a different threshold.

In the book we present a recipe for the application of the method that involves the likelihood ratio identity. This recipe is executed in a series of steps, and it is concluded by producing an approximation for the tail probability that we analyze. Unlike the baking of real cakes, one need not follow the proposed steps meticulously in order to avoid disasters. On the contrary, these steps are only guidelines and are not necessarily optimal in all scenarios. Still, we find them useful.

The first step involves the identification of the large deviation rate. The method itself produces refinements to the first-order approximation that is produced by this rate. Frequently, one may find the large deviation rate by the maximization of the marginal probabilities $\{P(\sup_{t \in T} X_t \geq x) : t \in T\}$ over the collection T.

A large deviation rate is associated with a collection of values in the parameter set. Preparation towards the application of the likelihood ratio identity may involve the identification of a subset of parameter values that are most likely to contain the maximizing value and restricting the analysis only to that sub-region.

In the case where the parameter space is continuous one may consider another preparation step in which the maximization is restricted to a dense, but discrete, sub-collection. Although the method can, and has been, implemented directly to a continuous parameter set there are some technical advantages to its implementation in the context of a discrete set of parameters.

After preparations one may invoke the likelihood ratio identity. The outcome of this step is a presentation of the probability, under the given distribution, that the maximum of the field exceeds a threshold in terms of a sum of expectations. The sum extends over the different values of the parameters. Each element in the sum is computed in the context of the alternative probability model specified by that value. The expectation involves a deterministic term that is associated with the large deviation rate and a product of two random terms, one measuring the global behavior of the field in the context of maximization and the other measuring local fluctuations.

The localization theorem, the subsequent step, applies a local limit theorem to the global term in order to prove the asymptotic independence between the effect of the global term and the effect of local fluctuations. In the examples that are considered the local limit theorem emerges as a refinement of the central limit theorem. Consequently, the given approach is more natural in problems where the global term obeys a central limit theorem and converges to the normal distribution. The outcome of this step is an approximation of each parameter-specific expectation in the representation by the product of three factors: a factor associated with the large deviation rate, a factor associated with the density of the normal limit of the global term, and a factor that measures local fluctuations. The integrated approximation of the tail probability is obtained by the summation of these products over the collection of parameter values.

The method is employed in settings consistent with large deviation formulation. Accordingly, the probability $P(\sup_{t \in T} X_t \geq x)$ converges to zero when the threshold x diverges to infinity. In other applications, when the parameter set T is increasing fast enough as a function of x, the probability may be converging to a positive constant. In such a case, Poisson-type approximations may be applied in order to extend the approximation obtained in large deviation settings to the setting of non-vanishing probabilities. Convergence in distribution that emerges from Poisson approximations, in conjunction with statement of uniform integrability, may be used in order to approximate functionals associated with extremes that involve expectation.

1.3   Gaussian and asymptotically Gaussian random fields

The random field is a collection of random variables $\{X_t : t \in T\}$ with a joint distribution. The joint distribution is uniquely specified in terms of the finite-dimensional joint distribution of the random vector $\{X_t : t = t_1, t_2 \ldots, t_k\}$, for any finite sub-collection of parameter values $t_1, \ldots, t_k \in T$. In the special case where these finite-dimensional joint distributions are all Gaussian we say that the field is a Gaussian random field.

The joint distribution of a Gaussian random vector, the multinormal distribution, is a function only of the vector of expectations and the matrix of variances and covariances. As a conclusion we get that the distribution of a Gaussian random field is fully specified in terms of the expectation function: the expectation $E(X_t)$, for each $t \in T$, and the variance-covariance function: the covariance $\text{Cov}(X_t, X_s)$, for any pair $(t, s) \in T \times T$. The distribution of the maximum of the Gaussian field is influenced both by the deterministic component of the field, namely the expectation, and by the variability, which is determined by the covariance function.

The theory that deals with the investigation of extreme values in Gaussian fields is highly developed. The role of the expectation, and the more delicate role of the covariance structure, in the determination of the distributions of such extremes is well understood. Extremely accurate asymptotic approximations of the distribution of these extremes exist for some subfamilies of Gaussian random fields. Good asymptotic approximations exist for other subfamilies.

In the next chapter we will present the two main tools for analyzing the distribution of extremes in Gaussian fields. One tool is based on the computation of the expectation of the Euler characteristic of the excursion set. This tool is applicable when the realizations of the random field are differentiable and is extremely accurate. The other, more general approach, is known as the double-sum method. It is not as accurate as the first tool but it may apply in situations where the realizations of the field are not smooth.

The smoothness, or lack thereof, of the realizations of a Gaussian random field is determined by the covariance function in general, and the smoothness of this function in the vicinity of the diagonal $\{(t, t) : t \in T\}$ in particular. Consequently, the answer to a question regarding which of the tools can be used is related to the ability to take partial derivatives of the covariance function at the points of the diagonal. Basically, if second-order partial derivatives exist then the more accurate method of the Euler characteristic may be applied. Otherwise, one should refer to the more general double-sum method.

Our method is more like the double-sum method in the sense that it may be applied under more general conditions, paying for the generality of the application in terms of accuracy of the approximation. It has a further advantage that it can be applied in settings where the random field is not Gaussian, although it should obey a local limit theorem that emerges from the central limit theorem.

The central limit theorem deals with the convergence of sums of random elements to a Gaussian element. If, for example, the elements are independent fields then the resulting limit is a Gaussian field. A central limit theorem in the context of random field relies, typically, on the convergence of the finite-dimensional distribution of the field to a Gaussian limit and on a tightness property. The role of tightness is to ensure that distribution of the field, along the process of convergence, may be approximated uniformly well using a finite and fixed collection of parameter points.

A tempting approach for dealing with the distribution of the extremes in a non-Gaussian setting, in the case where the field in question belongs to a sequence that converges to a Gaussian limit, is to apply the approximation of the distribution of the maximum to the Gaussian limit distribution. In this approach one separates between the convergence of the field to the Gaussian limit, which is carried out first, and the convergence of the tail distribution of the field to zero, which is assessed after the first convergence took place. Tempting as it may be, this approach may produce misleading outcomes. The reason for this is that the central limit theorem, as the name suggests, deals with the central part of the distribution, not with the extreme tail of the distribution. There may very well exist a big difference between the tail behavior of the original field and the tail behavior of a Gaussian field with the same expectation and the same covariance function as the original field. However, this difference is washed away by the central limit theorem. A better approach is to deal directly with the distribution of extreme values of the original field itself and assess its asymptotic behavior.

In the approach outlined in the previous section the probability of the maximum of the field is presented as a sum of terms. These terms are composed of a determinist factor that relates to the large deviation rate, a factor that relates to the contribution of the global term, and a factor that measures the contribution of local perturbations. Separating out the effect of large deviation guarantees an honest assessment of the extreme tail distribution. The central limit theorem plays a part in the derivation of the contribution of the global term and, in some cases, in the contribution of the local fluctuations.

The part played by the central limit theorem in the assessment of the contribution of the global term is in the form of a local limit theorem. A local limit theorem deals with the probability that a statistic, typically a statistic produced by taking a sum, obtains values from an interval of fixed width. The statistic has a variance that goes to infinity. Consequently, the probability of belonging to the interval goes to 0 at a rate proportional to the standard deviation of the statistic. An accurate assessment of the rate by which the probability converges to zero may be deduced from a central limit theorem that involves a higher order expansion of the approximation error. A famous theorem of that sort is the Berry–Esseen theorem. An important point to make is that the distribution of the global term is assessed in the context of an appropriate alternative distribution, not the original null distribution. Consequently, convergence may hold for the selected alternative distribution even if it does not exist for the original null distribution.

The method relies on a statement regarding the joint limit distribution of the global term and a local field that is derived by local deviations of the field. The requirement is a local limit for the global term and asymptotic independence between the global term and local deviations. The local deviations are not required to converge to a Gaussian limit. This requirement is much less than the requirement that the field converges to a Gaussian limit. In the particular important case where the field does converge to a Gaussian limit, for example when the field is a sum of independent fields, the factor in the approximation that is associated with the local fluctuation is the same as the factor that emerges for Gaussian fields (still, the factor that is associated with large deviation may be different). In other cases, the factor that is associated with local fluctuations may differ from the factors encountered in the Gaussian setting.

1.4 Applications

This book targets people with an interest in statistics and probability as branches of applied mathematics. As such, it will not do an honest job if it does not demonstrate the applicability of the theory to 'real life situations'. 'Application of mathematical theory' is, to some extent, an oxymoron. Typically, what is presented as applications are as abstract and removed from physical reality as the theory that it serves to demonstrate and justify. To justify the relation between an application and real life the mathematician tells a story that portrays the application as something that actual practitioners, people that do real science or put their money (or other people's money) at real risk, care about. These stories can fool outsiders but not people who actually know the details.

This description is true also for the applications given in this book. I am not an insider in either of the topics that appear in the second part of the book. In some cases, for example in genetics, I can say that I know people who know actual practitioners. In other cases I have only my imagination and what I read in Wikipedia to guide me.

Let us list the applications that will be presented in the book. In the first part of the book we will use two simple applications in order to demonstrate the solutions provided by the theory, both classical solutions that will be presented in Chapter 2, and the solution provided by our method. That solution will be mainly given in Chapter 3.

The first example is an example of sequential testing of hypotheses. In this example we will consider the case of two simple hypotheses and the application of the sequential probability ratio test for testing one hypothesis versus the other. In this example the parameter space is one-dimensional – the positive integers – and the random field is the process of a random walk. Tools that are applicable to random walks, in particular random walks stopped by a stopping time, can be used for the analysis of this example.

The second example is an example of a scanning statistic in the spirit of the example that was mentioned at the beginning of the chapter. Scanning statistics are used in order to detect rare signals in an environment contaminated by random noise. In the particular example we will consider a signal of the form of a region of elevated expectation in a linear environment with a Gaussian white noise. The elevated region is parameterized by the location of its center and by its width. The resulting field of scanning statistics is two-dimensional and Gaussian. We will consider cases where the field is smooth and a case where it is not. The classical tools for the analysis of Gaussian fields will be applied in Chapter 2 and our alternative approach will be used in Chapter 3. The Poisson approximation for the example of a scanning statistic and for a modified version of the sequential example is discussed in Chapter 4.

The second part of the book begins with a problem of intermediate difficulty. The task in that problem is to produce an approximation for the significance level of the Kolmogorov–Smirnov test and the Peacock test. Peacock's test is a generalization of the well known Kolmogorov–Smirnov nonparametric test of goodness-of-fit to higher dimensions. The Kolmogorov–Smirnov test compares the empirical distribution of a random variable to a theoretical distribution. Peacock's test, on the other hand, compares the empirical distribution of a random multivariate vector to its theoretical multivariate distribution. The analysis of the significance level of the Kolmogorov–Smirnov test is a classical exercise in many advanced statistics textbooks. Knowledge about the multivariate version of the test is not so wide spread. In Chapter 6 we give an alternative asymptotic derivation of the result for the Kolmogorov–Smirnov test on the basis of our approach. We then show how little effort is needed in order to extend the analysis to the multivariate setting.

More complex applications appear in the rest of the second part of the book. The applications presented are a reflection of the projects that I was involved with in recent years and in which the method was used.

The first application involves scanning for DNA copy number variations. Most of the genetic material in somatic cells comes in two copies, one originating from the mother and the other from the father. Occasionally, a segment of the DNA may be missing or may have multiple copies resulting in a copy number different than 2. A scanning statistic for detecting such intervals using genetic measurements generated from a sample can be constructed. The resulting scanning problem is not unlike the simple example that is used in the first part of the book. However, the produced two-dimensional field is not Gaussian, although it is asymptotically so. This problem is analyzed in Chapter 7.

The second example combines a scanning statistic with a sequential tool for change-point detection. This example involves a scenario in which an image is scanned for a signal of a specific structure. The specifications of the signal are not known nor is the time in which it will appear, if at all. The goal is to identify the emergence of the signal as fast as possible after it appeared, but not to do so before it did. The noise involved is Gaussian but the statistics that are used, and hence the associated random field, are not. The investigation of the properties of an appropriate stopping time that may be used to detect the emergence of a signal is carried out in Chapter 8.

The third application is discussed in Chapter 9. The issue in that application is the design of a buffer that is large enough to store packets waiting to be transmitted in an outgoing communication line. These packets arrive from a large number of independent sources and the outgoing line is of a fixed bandwidth. A simple model will characterize an incoming source by the distribution of the duration of time that is active, the 'on' period, and the distribution of time that it is idle, the 'off' period. The size of the required buffer can be associated with a level and the probability of a buffer overflow can be associated with the probability that a random field associated with the sum of the on–off processes exceeds the level. Apart from the fact that the field is not normally distributed, but only asymptotically so, it is also the case that the characteristic behavior of local fluctuations in this case differs from the characterizations in the other examples. This local behavior is specified by a parameter we denote by α that may take a value between 0 and 2 in general, and between 1 and 2 in the specific case. The local behavior is characterized by constants known as Pickands' constants that depend on this parameter α.

The Pickands' constants emerged as part of the development of the classical double-sum method for analyzing Gaussian fields and were defined a long time ago. However, their numerical evaluation remained an elusive problem and only crude upper and lower bounds to their value existed. In Chapter 10 we will explain how the representation that emerges from the method can be used in order to evaluate the constants efficiently. With this last application we conclude the book.

2

Basic examples

2.1 Introduction

In this chapter we give an informal description of the theory and some of the available tools for the approximation of the probability that a random field exceeds a high threshold. The theory is described in the context of two seemingly unrelated problems: a problem of sequential testing of hypotheses and a problem of scanning for a rare signal in a noisy Gaussian environment. Different tools will be applied to the different examples, tools that exploit the unique characteristics of the problems. The characteristic for the first problem is the fact that the set of parameters that defines the field is linearly ordered. The important feature for the second example is the fact that the field is Gaussian.

At the end of this chapter we list some other methods that appeared in the literature and have relevance to the problem at hand.

In the next chapter we will apply a different method to both problems. The similarity between the two examples will be revealed as a consequence.

2.2 A power-one sequential test

Imagine a dynamic setting where one is allowed to observe a system until enough information has been gathered in order to reach a conclusion. The issue at hand is to construct an efficient decision rule that can serve as a guide in making the decision to stop the process of observing the system and taking an appropriate action.

Such dynamic settings emerge for example in clinical trials, where the concern is to make a decision regarding the relative efficacy of a new medical treatment before releasing it to the market. Ethical consideration proposes a sequential trial design in which new subjects are not treated with the new procedure once enough evidence points to the new treatment as being inferior to other remedies. On the

Extremes in Random Fields: A Theory and its Applications, First Edition. Benjamin Yakir.
© 2013 by Higher Education Press. All rights reserved. Published 2013 by John Wiley & Sons, Ltd.

other hand, if the new treatment turns out to be superior then marketing it to the general public should not be delayed.

Another situation where sequential designs are useful is in industrial quality control. The quality of the manufactured products is monitored continuously. An alarm is set if there is accumulating evidence indicating a deterioration in the quality of the products. Quality concerns should be balanced against considerations of efficiency of production. Hence, one should make sure that the monitoring procedure does not produce false alarms too frequently.

In order to understand some of the probabilistic issues involved, let us consider a simplified version of the problem. Denote by $X_1, X_2, X_3, \ldots$ the sequence of observations that one may potentially make on the system. The statistical procedure is composed of a stopping time N, namely an integer-valued random variable with the property that, for each integer n, $\{N = n\}$ is a function of the first n observations $X_1, \ldots, X_n$, and a decision rule that is based on these same observations. Assume that the observations are independent and identically distribution. The exact form of the density of an observation is either $H_0 : X \sim f$ or $H_1 : X \sim g$, for two distinct density functions f and g that share the same support. A priori it is unknown which of the two densities is in effect and it is the role of the decision rule to make that determination on the basis of incoming data.

Considerations of statistical theory guide one to use likelihood ratios as statistics for testing a hypothesis. Equivalently, one may use the log of the likelihood ratio. In the presence of n observations this statistic becomes:

$$\ell_n = \sum_{i=1}^{n} \log \{g(X_i)/f(X_i)\} .$$

The increments of the sum are independent and identically distributed. If follows from Jensen's inequality (see Theorem A.12) that the expectation of an increment is negative if the actual density of an observation is f and it is positive if the actual density is g. Large values of this statistic support the alternative hypothesis H_1 whereas negative values of the statistic are consistent with the null hypothesis H_0.

An example of the path of the log-likelihood process, truncated after 40 observations, is plotted in Figure 2.1. The process has a negative drift, that is marked in the plot by a broken line.

A classical procedure, the sequential probability ratio test, uses these statistics as a monitoring sequence and continues sampling as long as the sequence is between prespecified upper and lower boundaries. Once a boundary is crossed the stopping time is activated and a decision made. If the upper boundary was the boundary that was crossed then the decision is to reject the null hypothesis and to accept the alternative. However, if the lower boundary was crossed first then the null hypothesis is not rejected.

A special form of the sequential probability ratio test is when the lower boundary is absent. For such a procedure, if the actual density is g then it is a

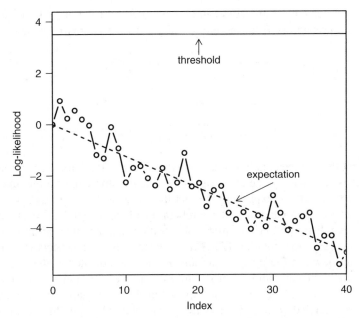

Figure 2.1 An example of the path of the log-likelihood process under the null hypothesis. The threshold is marked as a solid line. A broken line indicates the drift of the process.

corollary of the law of large numbers that the upper bound will be eventually crossed with certainty. Consequently, the probability of correctly rejecting the null hypothesis, the statistical power of the test, is equal to 1. Our concern in the context of this procedure will be to evaluate its significance level, i.e., the probability that it ever crosses the boundary when the density of the observations is from the null distribution f.

The power-one sequential probability test is associated with an upper boundary x. Denote the stopping time of the test by N_x, where:

$$N_x = \inf\{n : \ell_n \geq x\},$$

and $N_x = \infty$ if no such n exist. The significance level of the test is the probability $P(N_x < \infty)$ that N_x is finite, where the probability is computed under the null distribution that is governed by f. The threshold is marked as a solid line in Figure 2.1.

We can compute this probability by splitting the event among the disjoint events associated with stopping at particular times:

$$P(N_x < \infty) = \sum_{n=1}^{\infty} P(N_x = n).$$

Each of the elements in the sum can be computed by changing the density that governs the probability from the null density f to the alternative density g:

$$P(N_x = n) = \int_{\{N_x = n\}} f(x_1) \cdots f(x_n) dx_1 \cdots dx_n$$

$$= \int_{\{N_x = n\}} e^{-\ell_n} g(x_1) \cdots g(x_n) dx_1 \cdots dx_n$$

$$= E_g \left[e^{-\ell_n} ; N_x = n \right].$$

Notice that the last expectation is computed under the distribution governed by g. This fact is denoted by the subscript g attached to the expectation functional. Notice also that we are exploiting the fact that N_x is a stopping time. As a result of this fact the computation of the probability of the event $\{N_x = n\}$ involves only the joint distribution of the first n observations and no other. Finally, observe that we are using the notation $E[Y; A]$ to indicate that the expectation involves the product between the random variable Y and the indicator of the event A.

The 'trick' of computing a probability of an event formulated in some distribution by reformulating it as an expectation in the context of an alternative distribution is called *the likelihood ratio identity*. We will apply this trick over and over again throughout the book. In the current sequential setting one may observe that over the event $\{N_x = n\}$ we may write the log-likelihood ratio in the form $\ell_n = \ell_{N_x}$. Consequently, when substituting the probabilities in the sum by their representation as expectations, we get the sequential likelihood ratio identity:

$$P(N_x < \infty) = \sum_{n=1}^{\infty} E_g \left[e^{-\ell_{N_x}} ; N_x = n \right] = E_g \left[e^{-\ell_{N_x}} ; N_x < \infty \right].$$

For the last equality we exploited once more the fact that the events under consideration are disjoint and form a partition of the event $\{N_x < \infty\}$. Consequently, using the fact that under the alternative distribution $P_g(N_x < \infty) = 1$, we may express the significance level of the test in the form:

$$P(N_x < \infty) = E_g \left[e^{-\ell_{N_x}} \right] = e^{-x} E_g \left[e^{-(\ell_{N_x} - x)} \right]. \tag{2.1}$$

For illustration consult Figure 2.2. Following the likelihood ratio identity, the paths are generated under the alternative distribution that has a positive drift. An example of such a plot is given in the figure. The drift will guarantee that the threshold is crossed at some point. The stopping time corresponds to the first time it happens. Specifically, in the given figure $N_x = 30$. The overshoot, the discrepancy between the value of the process at N_x and the threshold, is marked on the plot.

We are interested in the probability that is given in (2.1) for large values of x. From the representation we may see that this probability converges to zero at an exponential rate. This rate is multiplied by a term that is given in the form of an expectation. What can be said about this expectation when x goes to infinity?

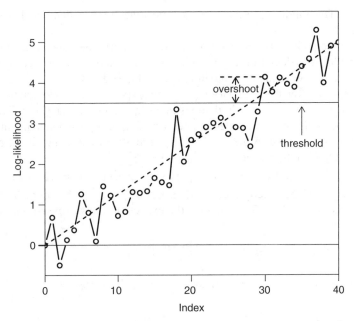

Figure 2.2   An example of a path of the log-likelihood process under the alternative distribution. The threshold is marked as a solid line. A broken line indicates the drift of the process. The stopping time is activated when the process crosses the threshold, n = 30 in the figure. The overshoot above the boundary is indicated as well.

The classical method for analyzing the given expectation is to apply renewal theory. Renewal theory deals with the process of partial sums of independent and positive random variables. The random variables symbolize cycle lengths. At the end of each cycle the system is restored and a new cycle is initiated. The theory deals with events that occur in a typical cycle.

The setting that led to (2.1) also involves partial sums of independent random variables. However, although the increments have a positive expectation under the alternative distribution, still they may obtain negative values and are thus not positive random variables. A remedy for this inconsistency with the assumption of renewal theory is to consider ladder variables as substitutes for the original variables. The ladder times τ_j are defined via:

$$\tau_1 = \inf\{n : \ell_n > 0\}, \quad \sum_{i=1}^{j} \tau_i = \inf\left\{n > \sum_{i=1}^{j-1} \tau_i : \ell_n > \ell_{\sum_{i=1}^{j-1} \tau_i}\right\},$$

for $j = 2, 3, \ldots$. The first ladder height is $Y_1 = \ell_{\tau_1}$ and the subsequent ladder heights are $Y_j = \ell_{\sum_{i=1}^{j} \tau_i} - \ell_{\sum_{i=1}^{j-1} \tau_i}$, considered as the periods between cycle renewals. The random variables Y_j are positive, independent, and identically distributed.

The same path that was given in Figure 2.2 is presented again in Figure 2.3. The cumulative ladder times are marked on the x-axis and they correspond to times when a new sequential maximum was obtained.

The excess over the boundary $\ell_{N_x} - x$ that appears in the expectation on the right-hand side of (2.1) and is demonstrated in Figure 2.2 can be reformulated in terms of ladder heights. Indeed, since crossing an upper boundary must correspond to an increase in the level of the partial sum and hence must coincide with a ladder height we get that: $\ell_{N_x} - x = \sum_{i=1}^{J_x} Y_i - x$, where $J_x = \inf\{j : \sum_{i=1}^{j} Y_i \geq x\}$. Thereby, we may express the expectation in (2.1) in terms of the overshoot of the strictly increasing process $S_j = \sum_{i=1}^{j} Y_i$, immediately after crossing the threshold x. To conclude, the expectation of the overshoot satisfies:

$$\mathrm{E}_g\left[e^{-(\ell_{N_x}-x)}\right] = \mathrm{E}_g\left[e^{-(S_{J_x}-x)}\right] ,$$

where S_j is the process of partial sums of ladder variables.

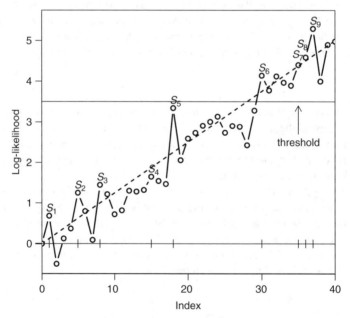

Figure 2.3   An example of a path of the log-likelihood process under the alternative distribution. The times of increase of the process, the ladder times, are marked as ticks on the x-axis. The cumulative ladder heights are indicated above the respective positions. The threshold is crossed at the end of the sixth cycle, so $J_x = 6$.

The cumulative ladder times S_j are marked in Figure 2.3 above respective sequential maxima. The maximum that led to the first crossing of the threshold is S_6. Correspondingly, $J_x = 6$. Clearly, $\ell_{30} = S_6$.

The advantage of considering partial sums of positive random variables is the fact that they form a strictly increasing sequence. Consequently, if the process is below the threshold at time j then we are assured that the process did not cross the threshold before that time. As a result, we may write the event associated with crossing at a given time in terms of the partial sum immediately before that time and the independent increment that follows:

$$\{J_x = j\} = \{S_{j-1} < x, S_j \geq x\} = \{S_{j-1} < x, Y_j \geq x - S_{j-1}\}.$$

(In order to be consistent with the notation when $j = 1$ we define $S_0 = 0$).

Again, we compute an expectation by partitioning it among disjoint events. Currently, the partition is with respect to the different values of the stopping time J_x. Moreover, we compute the value of the integral itself by conditioning on the value of the partial sum just before stopping and obtain:

$$\mathrm{E}_g \left[e^{-(S_{J_x} - x)} \right] = \sum_{j=1}^{\infty} \int_0^x \mathrm{E}_g \left[e^{-(Y_j - (x-u))}, Y_j \geq x - u \right] \mathrm{P}\left(S_{j-1} = u \right) du .$$

Notice that we used independence in order to replace the conditional expectation with respect to Y_j, given the value of S_{j-1}, by an expectation with respect to the marginal distribution of that random variable. Our notation assumes that S_j has a density. As a matter of fact, the argument does not really need such a requirement. Still, to keep notation simple, we will maintain this assumption.

Finally, after the change of variable $v = x - u$ and interchanging between summation and integration we get:

$$= \int_0^x \mathrm{E}_g \left[e^{-(Y_1 - v)}, Y_1 \geq v \right] \left\{ \sum_{j=0}^{\infty} \mathrm{P}(S_j = x - v) \right\} dv$$

$$= \int_0^x \mathrm{E}_g \left[e^{-(Y_1 - v)}, Y_1 \geq v \right] h(x - v) dv , \tag{2.2}$$

where $H(x) = 1 + \sum_{n=1}^{\infty} \mathrm{P}(S_n \leq x)$ is the renewal measure, defined for $x \geq 0$, and $h = H'$ is the density of this measure.

The analysis up until this point was relatively straightforward. Unfortunately, the subsequent analysis becomes much more involved and complex in details, although the principles are simple. We will not give a precise derivation of this section's example in the following. Instead, we will describe the principles and give an outline of the proof. Details can be found in [1], Chapter VIII, and the references therein.

Basically, the remainder of the derivation is an application of the renewal theorem in order to obtain the limit, as $x \to \infty$, of the integral in (2.2). This limit is given in terms of the distribution of the first ladder variable Y_1. A relation between the distribution of a ladder variables and the distribution of partial sums of the original increments is used in order to produce a formula for the limit in terms of the partial sums ℓ_n.

The renewal theorem is a type of a local limit theorem, for large values of x, of the renewal measure $H(x)$. Consider an interval of the form $[x - v, x]$ of length v. The key observation that is formulated by the theorem is that for large values of x the initial value of the process will have only a diminishing effect. Therefore, one cannot identify any particular region in the interval where renewals are more likely to occur compared with other regions. The consequence is that in the limit, as $x \to \infty$, the density is flat over the interval.

The theory distinguishes between the case where the values of Y_1 reside on an arithmetic lattice, in which case the measure H must also be concentrated on this lattice, and the non-arithmetic situation where no such lattice exists. We will be interested here only in the non-arithmetic case. The conclusion of the theorem in the non-arithmetic case is that the density $h(x - v)$ can be approximated by the constant $1/E(Y_1)$, for all values of v such that $x - v$ is still large.

Consider the integral in (2.2). It depends on x through the upper limit and through the evaluation of the renewal density at $x - v$, for $v \leq x$. Under appropriate regularity conditions on the tail of Y_1 one may make sure that the expectation within the integral, taken as a function of v, converges to zero fast enough for large values of v. Consequently, the evaluation of the integral will depend only on the value of v in a bounded interval. Replacing $h(x - v)$ by its approximation and letting the upper limit of the integral go to infinity produces the limit of (2.2):

$$\int_0^\infty E_g\left[e^{-(Y_1-v)}, Y_1 \geq v\right] \frac{dv}{E_g(Y_1)} = \frac{1 - E_g(e^{-Y_1})}{E_g(Y_1)}.$$

The equality follows from presenting the expectation as an integral and changing the order of the expectation integral with the outer integral.

Recall that the random variable Y_1 is the first ladder height. This random variable is associated with the first ladder time $\tau_1 = \tau_+ = \inf\{n : \ell_n > 0\}$ via the relation $Y_1 = \ell_{\tau_+}$. In other words, the variable corresponds to the random walk of log-likelihood ratios, stopped by an appropriate stopping time. In order to compute the expectation of a stopped random walk one may apply Wald's identity. This identity, which is straightforward to prove, states that the expectation of a stopped random walk is equal to the product of the expectation of an increment times the expectation of the stopping time (provided that both expectations are finite). Specifically, in our case we get that $E_g(\ell_{\tau_+}) = E_g(\ell_1)E_g(\tau_+)$. Plugging this identity in the quantity we seek to evaluate gives:

$$\frac{1 - E_g(e^{-Y_1})}{E_g(Y_1)} = \frac{1 - E_g(e^{-\ell_{\tau_+}})}{E_g(\ell_{\tau_+})} = \frac{1 - P(\tau_+ < \infty)}{E_g(\ell_1)E_g(\tau_+)}.$$

The value of the numerator in the last equality follows the application of the likelihood ratio identity.

In order to give an expression for the expectation of the stopping time τ_+ one may use an alternative representation of its survival function in terms of the event that the random walk obtains a sequential minimum at a given time:

$$P_g(\tau_+ > n) = P_g\left(\max_{0 \le k \le n} \ell_k \le 0\right) = P_g\left(\max_{0 \le k \le n}(\ell_n - \ell_k) \le 0\right)$$

$$= P_g\left(\ell_n = \min_{0 \le k \le n} \ell_k\right).$$

The validity of the second equation stems from the fact that the partial sums $\ell_n - \ell_k$ constructed in the reverse order have the same joint distribution as the original partial sums.

The expectation is produced by the summation of the survival probabilities. The sum yields:

$$E_g(\tau_+) = \sum_{n=0}^{\infty} P_g(\tau_+ > n) = E_g\left(\sum_{n=0}^{\infty} 1_{\{\ell_n = \min_{0 \le k \le n} \ell_k\}}\right).$$

The infinite sum is a count of the number of times that the random walk obtains a sequential minimum. The same count may be constructed by the aid of a new sequence of stopping times: the times between consecutive minima. For example, the first such time is: $\tau_- = \inf\{n \ge 1 : \ell_n \le 0\} = \tau_-^{(1)}$. Subsequent times $\tau_-^{(j)}$ are defined in a similar fashion, in parallel to the previous definition of the stopping times $\tau_j = \tau_+^{(j)}$. The count in the infinite sum terminates once a stopping time $\tau_-^{(j)}$ obtains an infinite value. The stopping times are independent and identically distributed. Consequently, the distribution of the total counts of sequential minima is geometric and therefore: $E_g(\tau_+) = 1/P_g(\tau_- = \infty)$. Adding this observation to the previous data leads to the representation:

$$\frac{1 - E_g(e^{-Y_1})}{E_g(Y_1)} = \frac{P_g(\tau_- = \infty)P(\tau_+ = \infty)}{E_g(\ell_1)}.$$

The final expression is obtained via representations of both probabilities in the numerator. These presentations are corollary of Theorem A.6 (obtained by letting $s \to 1$) and take the form:

$$-\log P(\tau_+ = \infty) = \sum_{n=1}^{\infty} \frac{1}{n} P(\ell_n > 0) \quad \text{and}$$

$$-\log P_g(\tau_- = \infty) = \sum_{n=1}^{\infty} \frac{1}{n} P_g(\ell_n \le 0).$$

Putting all things together we get the final expression for the limit:

$$\lim_{x \to \infty} e^x P(N_x < \infty) = \frac{\exp\left\{-\sum_{n=1}^{\infty} n^{-1}[P_g(\ell_n \le 0) + P(\ell_n > 0)]\right\}}{E_g(\ell_1)}. \tag{2.3}$$

Consider a specific case of testing in the normal shift model. Assume that the null density f is the standard normal density $\phi(x) = (2\pi)^{-\frac{1}{2}} \exp\left\{-\frac{1}{2}x^2\right\}$ and the alternative density is $g(x) = \phi(x - \mu)$, for some given $\mu > 0$. The log-likelihood of n observations is $\ell_n = \sum_{i=1}^{n}\{\mu X_i - \mu^2/2\}$. This random variable has the normal distribution with variance equal to $n\mu^2$, both under the null distribution and under the alternative. The expectation under the alternative is $n\mu^2/2$ and under the null it is the negative of the alternative expectation. From the symmetry and continuity of the normal distribution we get that the limit in this case is:

$$\lim_{x \to \infty} e^x P(N_x < \infty) = \frac{\exp\left\{-2\sum_{n=1}^{\infty} n^{-1}\Phi(-n^{\frac{1}{2}}\mu/2)\right\}}{\mu^2/2} = v(\mu),$$

where $\Phi(x)$ is the cumulative distribution function of the standard normal distribution.

The overshoot-correction function $v(\mu)$ has a reasonably nice looking representation. Unfortunately, this representation is not too convenient for numerical evaluation, especially when the value of μ is small, since in this case the convergence of the infinite sum is very slow. However, one may show that the limit value of the function when $\mu \to 0$ (in which case the Brownian motion is used for testing) is 1. Moreover, a Taylor expansion of the function near the origin is available (see, for example, [2]). For numerical purposes it was recommended in [3] to use the ad hoc formula:

$$v(\mu) \approx \frac{(2/\mu)(\Phi(\mu/2) - 0.5)}{(\mu/2)\Phi(\mu/2) + \phi(\mu/2)}. \tag{2.4}$$

2.3   A kernel-based scanning statistic

The second example we consider involves searching for signals in a noisy environment. The development of modern computers allows for the collection and analysis of large quantities of data. These data contain a lot of useful information. Unfortunately, the vast majority of the data is irrelevant for the specific task one seeks to carry out. Thus, a primary concern is to sort out the preciously few 'needles' of useful data that are hidden inside the pile of useless 'hay'.

Again, computers can come to our aid. Characteristics of the needles can be specified and an algorithm for the examination of the pile in order to efficiently identify objects with the given characteristics can be used for the task. Our goal in this section will not be the development of such algorithms. Instead, we will deal with general statistical characteristics of scanning that applies such algorithms.

Noisy environment corresponds to the situation where the hay in the pile has all types of shapes. Some of the shapes may share some of the characteristics of a useful needle, but in fact correspond to useless hay. As a result, a typical application of an algorithm will produce a collection of 'hits' or detections. Some of the hits may involve real needles and some not. These are, respectively, true and false detections. An important statistical concern is to make sure that the rate of hits that are unrelated to actual needles, i.e., false detections, is not too high.

In order to be concrete let us look at a simple example in the spirit of the example that was considered in Chapter 1. Assume that we are scanning a linear region. In general, the landscape is flat at a level we denote by 0. However, here and there there may be elevations of a specific shape. Assume that we can characterize the elevation in the functional norm: $\beta \cdot g((x-t)/h)/h^{\frac{1}{2}}$, where x is the location in the region of the data point we examine, t is the location of the center of the elevation, h is a parameter that characterizes the width of the elevated region, and β is a measure of the overall strength of the signal. The functional form of the function g is known and fixed. However, the parameters t and h and β are a priori unknown.

The noise in the environment is modeled as a Gaussian white noise dB_x, which is the continuous-time version of independent and identically distributed normal random variables. If the signal is present then the observed process is $dX_x = \beta \cdot g((x-t)/h)dx + dB_x$. If the signal is absent then the observed process is $dX_x = dB_x$, i.e., pure noise. The model of pure noise, the null hypothesis, corresponds to the case $H_0 : \beta = 0$. The alternative hypothesis that we would like to consider is $H_1 : \beta > 0$.

For given parameter values (t, h, β) we get, since the signal corresponds to a shift in the expectation in a Gaussian process, that the log-likelihood ratio statistic for testing these values against the null hypothesis of pure noise is:

$$\ell_{t,h,\beta} = \beta \int g((x-t)/h)/h^{\frac{1}{2}}dB_x - \frac{\beta^2}{2h} \int [g((x-t)/h)]^2 dx \ .$$

Statistical theory suggests to use the score statistic in order to test for weak signals. The score statistic is the derivative of the log-likelihood statistic, evaluated at the null value of the parameter. Specifically, if we consider the score with respect to β we get:

$$\frac{\partial}{\partial \beta}\ell_{t,h,\beta}|_{\beta=0} = \int g((x-t)/h)/h^{\frac{1}{2}}dB_x \ .$$

The expectation of this score statistic under the null distribution is 0 and the variance of the statistic under this distribution is $\int [g((x-t)/h)]^2 h^{-1}dx$.

Consequently, the standardized score statistic, for a given value of the parameter $\theta = (t, h)$, is:

$$Z_\theta = \frac{\int g((x - t)/h)/h^{\frac{1}{2}} dB_x}{\{\int [g((x - t)/h)]^2 h^{-1} dx\}^{\frac{1}{2}}} .$$

The parameter $\theta = (t, h)$ specifies the structural characteristics of a signal. In Figure 2.4 we may see an example of a signal, given as a solid line in the lower part of the figure, with a bell-shaped elevation on the right-hand side of the region. In reality, we do not get to see this line. Instead, we observe the points that are scattered about the solid line. These points, that are a composition of the signal and random noise, represent the data. The specific signal is associated with a value of the parameter θ, which corresponds to a point in the two-dimensional parameter space. For convenience, an arrow indicates the association between the signal and the parameter point. A statistic Z_1 is computed from the data on the basis of the assumed signal. The value of this statistic is marked on the plot above the point θ_1.

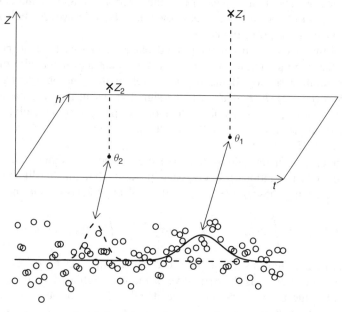

Figure 2.4   An example of the relation between the data, the parameters, and the scanning statistics. The data are plotted in the lower part of the figure along with two suggested models for the expectation. Model 1 is given as a solid line, and is parameterized by θ_1 and Model 2 is given as a broken line and is parameterized by θ_2. The scanning statistics Z_1 and Z_2 are computed from the data on the basis of Model 1 and Model 2, respectively. The resulting values of the statistic are plotted above their respective parameters.

Other possible signals are considered as well. The potential signal that is associated with the point θ_2 in the parameter space is presented as a broken line in the lower part of Figure 2.4. The value of the test statistic that is computed on the basis of this assumed signal, and is computed from the same data as before, is marked above the point θ_2 in the parameter space.

Large values of the standardized score statistic for a given parameter value indicate consistency between data and the presence of a signal in the form that is characterized by that value of the parameter. Hence, considering the score statistic as a function of the parameter, one's attention should be given to parameter values that produce high levels of the score statistic. A natural criteria is to set a threshold z and to single out all parameter values that produce statistic values at least as high.

Consider a given searching region T, i.e., a given subset of parameter values, and a threshold z. Typically, since true signals are rare, this region will not contain a signal. However, random fluctuation may result in relatively high levels of the score statistic and produce a false detection if they cross the threshold z. Consequently, in order to avoid a high rate of false detections one should set the threshold z high enough to be above most of the null fluctuations. In that context, one should be interested in the relation between the threshold z and the probability $P(\sup_{\theta \in T} Z_\theta \geq z)$, hoping to choose z such that this probability is small enough. This puts us again in a situation where the distribution of extreme values of a random field, a random field over a two-dimensional parameter space in this case, is relevant.

There are some important differences between the current random field and the one that appeared in the example of the previous section. The primary difference is that in the previous case the parameter space was one-dimensional. The one-dimensional Euclidean space is ordered. The tools of sequential analysis that were heavily exploited in the analysis of the previous example are not natural in spaces with no built-in ordering such as the two-dimensional parameter space of the current example. On the other hand, the current example involves a continuum of parameters whereas the parameter space of the previous example was discrete. Thereby, there is hope of being able to use in the analysis tools of differential and integral calculus of the field as a function of the parameters, tools that could not be used in the previous case.

There is a third difference that has importance in terms of the ability to carry out exact probabilistic computations. The current random field is a Gaussian random field, since the joint distribution of any finite collection of elements of the field is multivariate normal. The Gaussian distribution has unique properties that one can take advantage of in the development of asymptotic expansion of the tail distribution of extreme values of the random field.

In this section we will introduce two techniques that were developed in order to deal with maxima in a Gaussian random field. One of the techniques investigates, using tools of differential and integral geometry, the geometrical characteristics of elevated values of the random field Z_θ in order to produce a very accurate approximation of the probability in question. However, tools of differential and integral geometry assume the smoothness of the realization of the random field as a function of the parameters. Consequently, the price one pays for the accuracy in assessing the probability is in terms of a severe limitation of the range of applications in which the tool may be used. Specifically, it imposes smoothness assumptions on the realizations of the random field.

The other technique uses a more standard approach of dividing the search region into smaller subsets, assessing the probability of crossing in each such subset, and integrating the local approximations to produce a global approximation. This approach can be applied in a more general Gaussian field since it does not require smoothness. However, the accuracy of the resulting approximation is much less than for the geometrical approach. This other technique is referred to as the *double-sum method* due to the method that is used for establishing the validity of the integration of the local approximations. The technique that is being advocated here shares many similar features with the double-sum method. However, the two are not the same.

The geometrical method examines the excursion set, the subset of parameter values over which the random field obtains a value at least as large as the threshold:

$$A_z = \{\theta \in T : Z_\theta \geq z\} .$$

This set is a random set. It is not empty if, and only if, the maximal value of the field is larger than the threshold z:

$$P\left(\sup_{\theta \in T} Z_\theta \geq z\right) = P(A_z \neq \emptyset) .$$

Instead of investigating directly the content of the set A_z, which may be quite complex, the idea is to use a computable characteristic of the set as a substitute. Such characteristic is the Euler characteristic. Essentially, the Euler characteristic of the set is an integer value count of the number of subsets of a simple topological structure that make up the excursion set A_z, discounting double counts in the appropriate way. The approximation is based on the expectation of the given functional:

$$P\left(\sup_{\theta \in T} Z_\theta \geq z\right) \approx E(\varphi(A_z)) ,$$

where $\varphi(A_z)$ is the Euler characteristic of the excursion set.

The mathematical justification of the Euler characteristic method has two major components. The first component is to establish the validity of the approximation, namely to identify conditions that will assure that both the expectation of the Euler characteristic and the tail probability of the field share similar numerical values. The second component is to compute the expected Euler characteristic.

A full account of the mathematical theory that is associated with the method of the expected Euler characteristic of the excursion set can be found in the book by Adler and Taylor [4]. This book covers also the required background material, 120 or so pages on the properties of Gaussian fields and another 120 pages on basic and more advanced material in integral geometry that is needed for the proofs. Any attempt by me to give details of the proofs, or even to give a full description of the statement of the theorems, will be a disservice to the reader. The interested reader is referred to their book.

Instead of a real description of the approach we will give a superficial description of the two main theorems of the book and relate them to the discussion in the current text.

The first main theorem gives conditions that assure the validity of the approximation of the excursion probability by the expected Euler characteristic. This theorem, Theorem 14.3.3 in [4], states that for a smooth Gaussian process with variance equal to 1, where the Gaussian process is defined over a nice manifold T, one has:

$$\left| P\left(\sup_{\theta \in T} Z_\theta \geq z \right) - E(\varphi(A_z)) \right| < O\left(e^{-\frac{\alpha}{2}z^2} \right) ,$$

for some $\alpha > 1$. The Gaussian field is required to obey some extra conditions.

This theorem supports the praise expressed regarding the accuracy of the Euler characteristic approach. The leading term in the tail of the distribution of extreme values in such a random field is the marginal probability of exceeding the threshold $P(Z_\theta \geq z)$. When the variance is equal to 1 this probability is the survival function of the standard normal distribution, $1 - \Phi(z)$, which shares the same exponential term $e^{-\frac{1}{2}z^2}$ as the standard normal density. The actual probability is modified by a multiplicative term that is polynomial in z. The theorem establishes that the accuracy of the approximation of the probability by the expected Euler characteristic is super-exponential. As a corollary we get that the expected Euler characteristic captures the entire polynomial component correctly. In contrast, alternative approximations only share the same leading term of the polynomial.

The second theorem gives a computable expression for the Euler characteristic of the excursion set. The expression is given in terms of the Hermite polynomials:

$$H_n(x) = n! \sum_{j=0}^{\lfloor n/2 \rfloor} \frac{(-1)^j x^{n-2j}}{j!(n-2j)!2^j} , \quad n \geq 0 , \ x \in \mathbb{R} ,$$

and $H_{-1}(x) = \sqrt{2\pi}\,\Phi(x)e^{x^2/2}$, and in terms of a volume integral conducted on the manifold T and its boundaries, considered themselves as manifolds of lower dimensions:

Theorem 2.1 (Theorem 12.4.1 and Theorem 12.4.2 of [4]). *Let* Z_θ *be a centered, unit-variance Gaussian field on a d-dimensional, C^2 manifold T, and satisfying appropriate regularity conditions. Then*

$$E(\varphi(A_z)) = e^{-\frac{1}{2}z^2} \sum_{j=0}^{d} (2\pi)^{-(j+1)/2} \mathcal{L}_j(T) H_{j-1}(z) ,$$

where $\mathcal{L}_j(T)$ are the Lipschitz-Killing curvatures of T, calculated with respect to the metric induced by the field.

A notable property of the representation given in the theorem is the separation between the contribution of the marginal normal distribution to the polynomial part in the form of the Hermite polynomials and the spacial contribution of the field in the form of the Lipschitz-Killing curvatures $\mathcal{L}_j(T)$. These curvatures involve both the distribution of the random field and the geometrical properties of the manifold. The contribution of the distribution of the random fields is via the covariance matrix of the gradient field $\dot{Z}_\theta$, where the gradient is taken with respect to the parameter θ. The determinant of the covariance matrix $\Sigma_\theta = \mathrm{Var}(\dot{Z}_\theta)$ enters as a density that defines the measure used in the calculation of the curvatures.

The nice formulation of the theorem disguises the fact that the curvatures $\mathcal{L}_j(T)$ are rather difficult to compute. If the random field is stationary, in which case the density $|\Sigma_\theta|$ is constant, then one can produce an explicit formula for the simple rectangular manifold that is used in the example that we consider in this section. Unfortunately, the process in the example is not stationary so the computation of the Euler characteristic is more involved. On the other hand, if one gives up the attempt to compute the entire expression and is content with the computation of the highest order term only, the term then involves the highest power of z, then a simpler representation emerges.

The highest power appears in the Hermite polynomial with the highest index. In the two-dimensional example that we consider $j = d = 2$, which corresponds to $H_{2-1}(z) = H_1(z) = z$. The coefficient that is associated with this polynomial is $\mathcal{L}_j(T) = \int_T |\Sigma_\theta|^{\frac{1}{2}} d\theta$. The resulting approximation becomes

$$\mathrm{P}\left(\sup_{\theta \in T} Z_\theta \geq z\right) \sim z e^{-\frac{1}{2}z^2} (2\pi)^{-3/2} \int_T |\Sigma_\theta|^{\frac{1}{2}} d\theta . \tag{2.5}$$

Of course, with approximation (2.5) one may no longer claim a super-exponential accuracy.

Return to the example of a kernel-based scanning statistic that resulted in a centered and standardized Gaussian random field. The distribution of a Gaussian random field is fully characterized by the expectation and the covariance functions. The expectation function in this case is the constant at level 0. The covariance between two points $\theta = (t, h)$ and $\vartheta = (s, w)$ is given by:

$$\mathrm{Cov}(Z_\theta, Z_\vartheta) = \frac{\int g((x-t)/h)g((x-s)/w)\frac{dx}{\sqrt{hw}}}{\{\int [g((x-t)/h)]^2 \frac{dx}{h}\}^{\frac{1}{2}} \{\int [g((x-s)/w)]^2 \frac{dx}{w}\}^{\frac{1}{2}}} = \frac{\langle g_\theta, g_\vartheta \rangle}{\|g_\theta\| \, \|g_\vartheta\|},$$

where the inner product correspond to the inner product of the Hilbert space of square-integrable functions ($\langle g, f \rangle = \int g(x)f(x)dx$) and the norm in the denominator is the norm associated with this inner product. The notation for a signal that is associated with the parameter θ is: $g_\theta(x) = g((x-t)/h)/h^{\frac{1}{2}}$.

The validity of approximation (2.5) relies on the smoothness of the random field. Smoothness of a Gaussian random field is tightly linked to the smoothness of the covariance function. Specifically, we would like to explore the issue of smoothness in the context of our current example that involves integration of the function g_θ with respect to the Gaussian white noise. In order to simplify the discussion of smoothness, let us assume that the linear region of data points is unbounded and is composed of the entire real line. In such a case the norm of g_θ is constant as a function of θ and can be denoted $\|g\|$. The result is yet a simpler formula for the covariance function of the random field:

$$\mathrm{Cov}(Z_\theta, Z_\vartheta) = \langle g_\theta, g_\vartheta \rangle / \|g\|^2 . \tag{2.6}$$

The smoothness of the covariance function in (2.6) is related to the smoothness of the function g. If the function is smooth and the derivatives are integrable, $g(x) = \exp\{-x^2/2\}$ is an example, then one can express the derivatives of the covariance function in terms of the derivatives of the function g:

$$\frac{\partial}{\partial \vartheta}\mathrm{Cov}(Z_\theta, Z_\vartheta) = \langle g_\theta, \dot{g}_\vartheta \rangle / \|g\|^2 , \quad \left(\frac{\partial}{\partial \vartheta}\right)^2 \mathrm{Cov}(Z_\theta, Z_\vartheta) = \langle g_\theta, \ddot{g}_\vartheta \rangle / \|g\|^2 ,$$

etc., where we use the dot notation to express partial derivatives of the function $g_\vartheta(x)$ with respect to ϑ.

Let us now look more closely at the relation between the pathwise derivatives of the random field and the derivatives of the covariance function. An element of the field has the representation:

$$Z_\theta = \int g_\theta(x)dB_x / \|g\| .$$

The gradient with respect to θ of this element, for a smooth function g, is:

$$\dot{Z}_\theta = \int \dot{g}_\theta(x)dB_x / \|g\| ,$$

and it is defined for each θ in the interior of T. Likewise, the Hessian is $\ddot{Z}_\theta = \int \ddot{g}_\theta(x)dB_x / \|g\|$. Similarly, one can take higher derivatives of the random field, depending on the smoothness of g. Notice that all these derivatives are integrals of the Gaussian white noise, hence Gaussian. Moreover, the joint distribution of the original fields and its derivatives is multivariate normal.

Clearly, the expectation of the gradient vector is the zero vector. The variance-covariance matrix of the gradient is given by:

$$\Sigma_\theta = \mathrm{Var}(\dot{Z}_\theta) = \int [\dot{g}_\theta(x)][\dot{g}_\theta(x)]' dx / \|g\|^2 = \langle \dot{g}_\theta, \dot{g}_\theta \rangle / \|g\|^2 ,$$

where we interpret the expression $\langle \dot{g}_\theta, \dot{g}_\theta \rangle$ as the $d \times d$ matrix with components $\langle \dot{g}_\theta, \dot{g}_\theta \rangle_{ij} = \langle [\dot{g}_\theta]_i, [\dot{g}_\theta]_j \rangle$.

In particular, if $g(x) = \exp\{-x^2/2\}$ then $\|g\|^2 = \sqrt{\pi}$ and:

$$\dot{g}_\theta(x) = \begin{pmatrix} h^{-\frac{5}{2}}(x-t) \\ h^{-\frac{7}{2}}(x-t)^2 - \frac{1}{2}h^{-\frac{3}{2}} \end{pmatrix} e^{-\frac{1}{2h^2}(x-t)^2} .$$

Consequently, since:

$$\langle \dot{g}_\theta, \dot{g}_\theta \rangle_{11} = \int h^{-5}(x-t)^2 e^{-\frac{1}{h^2}(x-t)^2} dx = \frac{1}{h^2}\sqrt{\pi} ,$$

$$\langle \dot{g}_\theta, \dot{g}_\theta \rangle_{22} = \int \left\{ h^{-7}(x-t)^4 - h^{-5}(x-t)^2 + \frac{1}{4}h^{-3} \right\} e^{-\frac{1}{h^2}(x-t)^2} dx = \frac{1}{4h^2}\sqrt{\pi} ,$$

and $\langle \dot{g}_\theta, \dot{g}_\theta \rangle_{12} = 0$ we get that

$$\Sigma_\theta = \begin{pmatrix} h^{-2} & 0 \\ 0 & \frac{1}{4}h^{-2} \end{pmatrix} \quad \Rightarrow \quad |\Sigma_\theta|^{\frac{1}{2}} = \frac{1}{2}h^{-2} .$$

In the particular case where the parameter set is the rectangle $T = [t_0, t_1] \times [h_0, h_1]$ then approximation (2.5) specifies to:

$$P\left(\sup_{\theta \in T} Z_\theta \geq z \right) \sim z e^{-\frac{1}{2}z^2} (2\pi)^{-3/2} \cdot 0.5(t_1 - t_0)(1/h_0 - 1/h_1) . \qquad (2.7)$$

The entire argument falls apart if the kernel g is not a continuous function. For example, let $g(x)$ be the indicator of the interval $[-0.5, 0.5]$. Now $\|g\| = 1$ and

$$\text{Cov}(Z_\theta, Z_\vartheta) = \langle g_\theta, g_\vartheta \rangle = \frac{\min\left\{ t + \frac{h}{2}, s + \frac{w}{2} \right\} - \max\left\{ t - \frac{h}{2}, s - \frac{w}{2} \right\}}{\sqrt{hw}}$$

when the numerator is positive and $\text{Cov}(Z_\theta, Z_\vartheta) = 0$ otherwise. This function is equal to 1 when $\vartheta = \theta$ and it is continuous at this point, but it does not have derivatives there. Consequently, the realization of the random field does not have a gradient, although it can be selected to be continuous with probability 1. Approximation (2.5) no longer applies.

An illustration of a scanning statistic that is based on an indicator is given in Figure 2.5. The data in the figure are exactly the same data that were presented in Figure 2.4 in the context of a smooth kernel. However, the researcher is not aware of the fact that the model that generated the data emerged from a smooth kernel and proposed analyzing the data using shifting windows of different lengths, which corresponds to an indicator kernel. The parameter space is the same but the models are different. Consequently, the values of the scanning statistics are not the same and their distribution is different.

An alternative method for obtaining an approximation is the double-sum method. This is a general-propose method that can be used for Gaussian fields with a wide variety of covariance structures. We will use this method for the analysis of the last example involving a scanning statistic that uses a noncontinuous

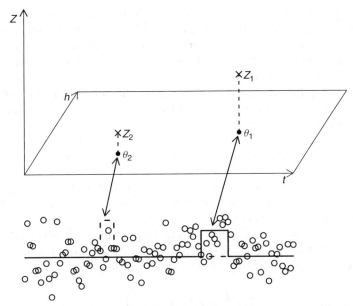

Figure 2.5   An example of the relation between the data, the parameters, and the scanning statistics when the kernel is an indicator. The data are plotted in the lower part of the figure along with two suggested models for the expectation. Model 1 is given as a solid line, and is parameterized by θ_1 and Model 2 is given as a broken line and is parameterized by θ_2. The scanning statistics Z_1 and Z_2 are computed from the data on the basis of Model 1 and Model 2, respectively. The resulting values of the statistic are plotted above their respective parameters.

kernel. The same method may be used in order to obtain the first-order approximation that was specified for the case of a smooth kernel.

A good source for learning about the double-sum method and its applications is [5]. Our description of the method will diverge slightly from the more general discussion that is presented there since our main aim is to emphasis the similarities and differences between the double-sum method and our method.

The basic argument in the double-sum method involves a partitioning of the parameter space into smaller regions. In each of the regions the field may or may not cross the threshold. The probability that the global maxima of the field is above the threshold, i.e., the probability of the union of events of crossing in a sub-region, is bounded from above by the sum of probabilities of crossing in a sub-region and bounded from below by the same sum, minus the double-sum of crossing simultaneously in a pair of distinct sub-regions. Gaussian arguments are used in order to obtain an approximation of the probability of crossing in a sub-region and thus an approximation of the sum. On the other hand, it is shown that the double-sum is of a smaller order compared with the sum, establishing the sum as the approximation.

In the given example it is convenient to change the parametrization for the definition of sub-regions. The scanning statistic in the current situation corresponds to integration of the data process over an interval of width h that is centered at t. Occasionally, we would like to characterize such intervals by their endpoints $\theta_1 = t - h/2$ and $\theta_2 = t + h/2$.

Fix $\theta = (\theta_1, \theta_2)$, the bottom-left corner of the sub-region. The sub-region involves small perturbations in the 'positive' direction of the endpoints about the given parameter value:

$$ T_\theta = \left\{ \vartheta = (\vartheta_1, \vartheta_2) : 0 \le \vartheta_1 - \theta_1 \le \tau/z^2, 0 \le \vartheta_2 - \theta_2 \le \tau/z^2 \right\} . $$

In the proof we fix τ and let $z \to \infty$, resulting in a shrinkage of the size of the sub-region and convergence to a limit. Subsequently, we increase τ in order to obtain the final approximation.

The critical computation corresponds to the probability of crossing the threshold z within the local region. We carry out this computation by conditioning on the value of Z_θ, which has a standard normal distribution:

$$ P\left(\max_{\vartheta \in T_\theta} Z_\vartheta \ge z \right) = E\left[P\left(\max_{\vartheta \in T_\theta} Z_\vartheta \ge z \mid Z_\theta \right) \right] $$

$$ = \frac{1}{\sqrt{2\pi}} \int e^{-\frac{1}{2}y^2} P\left(\max_{\vartheta \in T_\theta} Z_\vartheta \ge z \mid Z_\theta = y \right) dy . $$

Changing the variable to $x = -z(y - z)$ produces:

$$ = \frac{e^{-\frac{1}{2}z^2}}{\sqrt{2\pi}z} \int e^{x - \frac{1}{2}(\frac{x}{z})^2} P\left(\max_{\vartheta \in T_\theta} Z_\vartheta \ge z \mid Z_\theta = z - x/z \right) dx $$

$$ = \frac{e^{-\frac{1}{2}z^2}}{\sqrt{2\pi}z} \int e^{x - \frac{1}{2}(\frac{x}{z})^2} P\left(\max_{\vartheta \in T_\theta} z(Z_\vartheta - Z_\theta) \ge x \mid Z_\theta = z - x/z \right) dx . $$

This is a good place to pause for a while and examine the last expression. It involves the term $\phi(z)/z$, which is asymptotic to the survival function of the standard normal distribution, and an integral. The first term reflects the marginal distribution of the field elements. The integral corresponds to the contribution of the local field about θ. Elements of the local field are $z(Z_\vartheta - Z_\theta)$, considered in the conditional distribution given $Z_\theta = z - x/z$.

The local field, as $z \to \infty$, is subject to two negating influences that balance each other. On the one hand, shrinkage of the local region is shrinking the increments $Z_\vartheta - Z_\theta$. On the other hand, a multiplication of these increments by z is counterbalancing the effect of shrinkage. The rate of shrinkage is selected to assure convergence of the local field to a limit random field. In the limit, the integral becomes an appropriate functional of the limit of the local field.

Concrete computations may help clarify the reasons for selecting the local region. In the current parametrization we may write the covariance between two elements of the random field in the form:

$$\text{Cov}(Z_\theta, Z_\vartheta) = \frac{\theta_2/h - \vartheta_1/h}{\sqrt{1 + (\vartheta_2 - \theta_2)/h - (\vartheta_1 - \theta_1)/h}}$$

$$= \frac{1 - (\vartheta_1 - \theta_1)/h}{\sqrt{1 + (\vartheta_2 - \theta_2)/h - (\vartheta_1 - \theta_1)/h}} .$$

When $\vartheta = \theta$ the covariance is equal to 1. For other ϑ in the sub-region the covariance may be approximated by:

$$1 - \text{Cov}(Z_\theta, Z_\vartheta) \approx \frac{1}{2h}(\vartheta_1 - \theta_1) + \frac{1}{2h}(\vartheta_2 - \theta_2) = \frac{u_1}{2hz^2} + \frac{u_2}{2hz^2} ,$$

where $u_i = z^2(\vartheta_i - \theta_i)$, $i = 1, 2$. In general, for values of ϑ in the vicinity of θ, not necessary in the first quadrant with respect to it, we will find that $1 - \text{Cov}(Z_\theta, Z_\vartheta)$ is asymptotic to $[|u_1| + |u_2|]/[2hz^2]$.

The local random field $z(Z_\vartheta - Z_\theta)$, also in the conditional distribution given the value of Z_θ, is a Gaussian random field. As such, it is characterized by the expectation of elements and by the covariance structure. The conditional expectation is:

$$E(z(Z_\vartheta - Z_\theta)|Z_\theta = z - x/z) = -zZ_\theta(1 - \text{Cov}(Z_\vartheta, Z_\theta)) \approx -\frac{1 - \frac{x}{z^2}}{2h}(u_1 + u_2).$$

The conditional covariance does not depend on the specific value of Z_θ. One may obtain the conditional covariance between $z(Z_\vartheta - Z_\theta)$ and $z(Z_\eta - Z_\theta)$ by the examination of the conditional variance of the difference $z(Z_\vartheta - Z_\eta)$:

$$\text{Var}(z(Z_\vartheta - Z_\eta)|Z_\theta) = z^2\{2[1 - \text{Cov}(Z_\vartheta, Z_\eta)] - [\text{Cov}(Z_\vartheta, Z_\theta) - \text{Cov}(Z_\eta, Z_\theta)]^2\}.$$

The squared term inside the curly brackets is of order z^{-4} and will make a vanishingly small contribution to the covariance. Setting $\eta = \theta$ we get the variance of an increment:

$$\text{Var}(z(Z_\vartheta - Z_\theta)|Z_\theta) \approx 2z^2(1 - \text{Cov}(Z_\vartheta, Z_\theta)) \approx \frac{1}{h}(u_1 + u_2) .$$

A similar expression is obtained for the conditional variance of $z(Z_\eta - Z_\theta)$, with u_i replaced by $v_i = z^2(\eta_i - \theta_i)$, $i = 1, 2$. Combining these facts with the general expression for $1 - \text{Cov}(Z_\vartheta, Z_\eta)$ will give:

$$\text{Cov}(z(Z_\vartheta - Z_\theta), z(Z_\eta - Z_\theta)|Z_\theta) \approx \frac{1}{h}(u_1 \wedge v_1 + u_2 \wedge v_2) .$$

The covariance structure of a Brownian motion is given by $\text{Cov}(B_u, B_v) = u \wedge v$. We conclude that for each fixed x, and when $z \to \infty$, the expectation and the covariance structure of the local field converge to a limit expectation and covariance structures. The limit expectation and covariance structures, parameterized by the pair (u_1, u_2), is equal to that of a sum of a pair of independent Brownian motions with negative drifts:

$$\left\{ B^{(1)}_{u_1/h} - (1/2)(u_1/h) \right\} + \left\{ B^{(2)}_{u_2/h} - (1/2)(u_2/h) \right\} = \ell^{(1)}_{u_1/h} + \ell^{(2)}_{u_2/h}.$$

Storing the information given above, let us return to the task of evaluating the probability that the field crosses the threshold within the local region. As part of this evaluation we need to assess the convergence of the integral:

$$\int e^{x - \frac{1}{2}(\frac{x}{z})^2} P\left(\max_{\vartheta \in T_\theta} z(Z_\vartheta - Z_\theta) \geq x \mid Z_\theta = z - x/z \right) dx .$$

For each fix x, as $z \to \infty$, the exponent converges to e^x and the probability converges to the probability that the maximum of the limit field is above x. Consequently, provided that uniform integrability can be established, the integral converges to:

$$\int e^x P\left(\max_{0 \leq u_1 \leq \tau} \ell^{(1)}_{u_1/h} + \max_{0 \leq u_2 \leq \tau} \ell^{(2)}_{u_2/h} \geq x \right) dx = \{ \mathcal{H}(\tau/h) \}^2 ,$$

for

$$\mathcal{H}(t) = E\left(\max_{0 \leq u \leq t} e^{\ell_u} \right) ,$$

which is defined with respect to ℓ, a Brownian motion with an appropriate negative drift.

The uniform integrability requires a uniform bound on the tail of the maximum of a random field. Such a bound is provided for Gaussian random fields by Borell's inequality. This inequality states that the tail of the maxima is no more than twice the survival function of the normal distribution with expectation equal to the expectation of the maximum of the random field and with variance equal to the maximal variance of components of the field. Entropy considerations can be used to bound the expectation of the maximum and finish the proof that:

$$\sqrt{2\pi} z e^{\frac{1}{2}z^2} P\left(\max_{\vartheta \in T_\theta} Z_\vartheta \geq z \right) \to_{z \to \infty} \{ \mathcal{H}(\tau/h) \}^2 . \qquad (2.8)$$

In order to obtain an approximation for the distribution of the global maximum over the entire region $T = [t_0, t_1] \times [h_0, h_1]$ one splits the set into practically disjoint subsets T_{ij}. The linear transformation A maps the parametrization (t, h) to the new parametrization (θ_1, θ_2), where

$$A = \begin{pmatrix} 1 & -0.5 \\ 1 & 0.5 \end{pmatrix} .$$

In particular, $|A| = 1$, so this transformation preserves measure.

Consider the regular grid with span τ/z^2 over the (θ_1, θ_2) plane. The application of the inverse transformation A^{-1} to this grid splits the (t, h) plane into rhombuses. Let T_{ij} be any of the rhombuses that cover T, identified by its bottom corner. There is a total of $(t_2 - t_1)(h_2 - h_1)z^4/\tau^2$ such rhombuses of area τ^2/z^4 each. When we apply the asymptotic approximation to the crossing probabilities in each of these sub-regions we get the asymptotic upper bound:

$$
\begin{aligned}
P(\max_{\vartheta \in T} Z_\vartheta \ge z) &= P\left(\bigcup_{i,j} \left\{ \max_{\vartheta \in T_{ij}} Z_\vartheta \ge z \right\} \right) \\
&\le \sum_{i,j} P\left(\max_{A\vartheta \in AT_{ij}} Z_{A\vartheta} \ge z \right) \\
&\approx \frac{z^3 e^{-\frac{1}{2}z^2}}{\sqrt{2\pi}}(t_2 - t_1) \int_{h_0}^{h_1} \frac{1}{h^2}\left\{ \frac{\mathcal{H}(\tau/h)}{\tau/h} \right\}^2 dh .
\end{aligned}
\tag{2.9}
$$

The derivation of the asymptotic upper bound, which turns out to be the approximation itself, is completed by letting $\tau \to \infty$. To that end a likelihood ratio identity can again be applied. Indeed, we use the representation of $\mathcal{H}(t)$ in terms of an integral of an exponent times a survival function:

$$
\mathcal{H}(t) = E\left(\max_{0 \le u \le t} e^{\ell_u} \right) = \int_0^\infty e^x P\left(\max_{0 \le u \le t} \ell_u \ge x \right) dx + 1 .
$$

Notice that the integral for negative values of x is equal to 1 since the probability that the maximum is non-negative is equal to 1.

We would like to write the event in the probability in terms of a stopping time. Let $N_x = \inf\{u : \ell_u \ge x\}$ be the first time that the process ℓ_u reaches the level x. Stopping will have occurred by time t if, and only if, the maximum of the process in that interval of time is no less than x. Consequently, $\{\max_{0 \le u \le t} \ell_u \ge x\} = \{N_x \le t\}$.

We apply the likelihood ratio identity in order to compute the probability. Interestingly enough, the statistic ℓ_u is the log-likelihood ratio for testing $H_0 : \mu = 0$ versus $H_1 : \mu = 1$ on the basis of observing the Brownian motion over the interval of time $[0, u]$. Using the sequential likelihood ratio identity, which is justified by essentially the same argument as in the previous section, we get the representation:

$$
\int_0^\infty e^x P\left(N_x \le t \right) dx = \int_0^\infty e^x E_1\left(e^{-\ell_{N_x}} ; N_x \le t \right) dx .
$$

However, due to the continuity of the path of the Brownian motion, there is no overshoot involved. Upon stopping, over the event $\{N_x \le t\}$, we have that $\ell_{N_x} = x$. Consequently, we get that the integral is:

$$
= \int_0^\infty P_1\left(N_x \le t \right) dx = \int_0^\infty P_1\left(\max_{0 \le u \le t} \ell_u \ge x \right) dx = E_1\left(\max_{0 \le u \le t} \ell_u \right) .
$$

The last equality results from the standard representation of the expectation in terms of the integral of the survival function.

We may conclude that:

$$\mathcal{H}(t) = E_1 \left(\max_{0 \le u \le t} \ell_u \right) + 1 = t/2 + E_1 \left(\max_{0 \le u \le t} (\ell_u - \ell_t) \right) + 1 \,,$$

where $t/2$ is the expectation of ℓ_t under the alternative distribution that assigns the coefficient 1 to the expectation of the Brownian motion. This expectation is the leading term. In order to bound the second term observe that the joint distribution of the process $\ell_u - \ell_t = -(\ell_t - \ell_u)$, considered in reverse order for u ranging from t to 0, is the same as the null distribution of the original process. As a result:

$$0 \le E_1 \left(\max_{0 \le u \le t} (\ell_u - \ell_t) \right) = E \left(\max_{0 \le u \le t} \ell_u \right) \le E \left(\max_{0 \le u < \infty} \ell_u \right) = 1 \,.$$

The equality to 1 follows from the fact that $P(\max_{0 \le u < \infty} \ell_u \ge x) = e^{-x}$, which is the significance level of the power-one sequential probability ratio test when observing a Brownian motion. Again, no overshoot is involved.

Insert in (2.9) the fact that $\mathcal{H}(t)/t \to 0.5$, for $\tau \to \infty$, and obtain an asymptotic upper bound for the probability that the maximum of the field is larger than z. The expression for the upper bound becomes:

$$P \left(\max_{\vartheta \in T} Z_\vartheta \ge z \right) \sim z^3 e^{-\frac{1}{2}z^2} (2\pi)^{-\frac{1}{2}} \cdot (0.5)^2 (t_1 - t_0)(1/h_0 - 1/h_1) \,. \qquad (2.10)$$

This completes the easy part of the proof.

The more difficult part of the proof is to show that the asymptotic upper bound is tight. The double-sum method establishes that fact via the investigation of the sum of probabilities of simultaneous crossing in a pair of distinct sub-regions:

$$\sum_{(i,j)} \sum_{(k,l) \ne (i,j)} P \left(\max_{A\vartheta \in AT_{ij}} Z_{A\vartheta} \ge z, \ \max_{A\eta \in AT_{kl}} Z_{A\eta} \ge z \right) \,.$$

We will not present the proof that this double sum is of lower order. The interested reader may refer to [5]. However, we will mention the fact that the proof employs Slepian's inequality, which stochastically bounds the extreme value of a Gaussian field in terms of the extreme value of a Gaussian field with smaller correlation between components. This inequality is unique to the Gaussian case and does not generalize to non-Gaussian settings.

2.4   Other methods

In this chapter we introduced a method for analyzing the maximum of a one-dimensional random process and two methods for the analysis of Gaussian multi-dimensional fields. Here we mention other methods.

The history of analyzing one-dimensional processes is longer and resulted in the development of more methods. Typically, a method for a random field started its life as a method for a random process and then got upgraded. For example, one may find the roots of the geometric Euler characteristic approach dating back to Rice's formula for counting up-crosses in a smooth Gaussian process. Like the expected Euler characteristic, Rice's provides an approximation for the maxima of such a process.

An alternative geometric approach for analyzing smooth Gaussian fields is the tube method. This method relies on the representation of the Gaussian field as a sum of smooth functions using the Karhunen–Loève expansion. The coefficients in the expansion are independent standard normal variables. By a truncation of the expansion one obtains a finite collection of such variables that generates the randomness of the field. For each parameter value, the expansion becomes an inner product between a point in a manifold and the vector of a standard multivariate normal distribution. A multivariate standard normal vector is generated by the selection of a uniform direction on the unit sphere and the independent selection of the squared norm according to the chi-square distribution. The statement that the random field crosses a high threshold is translated to a statement about the volume of a tube about the manifold with a random radius that is determined by the distribution of the norm of the Gaussian vector. The expected value of the volume of the tube produces the approximation. The tube method is restricted to smooth Gaussian fields and in that context is less general than the Euler characteristic approach. Still, it has its merits. A less superficial description of the tube method can be found in [6].

Highly developed techniques for one-dimensional processes, such as renewal theory, martingale methods and stochastic calculus, are relevant as tools for the investigation of the distribution of maxima. These and more are highly developed for one-dimensional processes and are relevant as tools for the investigation of the distribution of maxima. It is hopeless to try to cover these vast areas in this short section. Generally speaking, these methods rely on the ordering of the parameter set. Sometimes this ordering is used in order to partition an event into disjoint events which are easier to handle. For example, in the sequential likelihood ratio argument an event was partitioned according to the values of a random time first reaching a threshold. Other partitions have been used successfully. For example, an analysis of the crossing probability for a random process can be carried out by considering the *last* time that the process went above the threshold. In some cases this partition is more convenient than the partition according to the *first* time that the threshold was crossed. As mentioned before, methods that are based on the ordering of events and the partitioning according to that order can be used in higher dimensions by imposing an ordering on the parameter space.

In Aldous' book [7], there is an intuitive idea for writing an expression for the probability of the maxima in terms of an expected size of some random sets. Instead of struggling with an explanation let me quote the author's explanation from the introduction in the book:

Let us try to say this idea in one paragraph.

(a) Problems about random extrema can often be translated into problems about sparse random sets in d = 1 dimensions.

(b) Sparse random sets often resemble i.i.d. random clumps thrown down randomly (i.e., centered at points of a Poisson process).

(c) The problem of interest reduces to estimating mean clump size.

(d) This mean clump size can be estimated by approximating the underlying random process locally by a simpler, known process for which explicit calculations are possible.

(Part (b) explains the name Poisson clumping heuristic).

We do not use in this book the Poisson clumping heuristic. Yet, this heuristic has some resemblance to and relationship with the material of Chapter 4 where we deal with the Poisson approximation. In that chapter we discuss a method for expanding the approximation from the regime where probabilities converge to 0 to the regime where they are not. That generalization involves a Poisson argument and is in the flavor of the Poisson clumping heuristic.

However, as far as we are concerned, the more difficult part in the analysis of the distribution of extremes in random fields is the computation of the rate of the Poisson distribution, a task which is carried out in Chapter 3 and is where most of the effort is invested in the other applications that are presented in the second part of the book. This part of the argument is comparable with the problem of computing the expected clump size. So, if one chooses, one can view this book as providing yet another method for computing the expected clump size and adding along the way five more examples to Aldous' 100 examples.

More seriously, in this book we propose a specific method for computing the Poisson rate of the number of clumps. Consequently, by reversing Aldous' argument one may come up with a method to approximate the expected clump size on the basis of the marginal probabilities for crossing the threshold, i.e., the rate of occurrences, divided by the Poisson rate of the number of clumps. In applications where the size of a typical clump is relevant this information may be useful.

3

Approximation of the local rate

3.1 Introduction

In the previous chapter we presented an array of methods that can be used in order to analyze the distribution of extremes in a random field. In this chapter we concentrate on yet another method, a method that is based on a transformation of the measure. The components of this method will be demonstrated in the context of the two examples that were introduced in the previous chapter.

The first example involved sequential testing of hypotheses. This example deals with the case where a stream of independent observations is accumulating. The aim is to test the null hypothesis that the density of the observations is f versus the alternative hypothesis that the density is g. The statistics that are used are the log-likelihood ratio statistics:

$$\ell_n = \sum_{i=1}^{n} \log\{g(X_i)/f(X_i)\} .$$

These statistics, in combination with a threshold x, define the test's stopping time: $N_x = \inf\{n : \ell_n \geq x\}$.

The significance level of the test, the probability under the null regime that the stopping time is finite, was analyzed in the previous chapter using tools of sequential analysis. In particular, a sequential likelihood ratio identity was applied. The resulting approximation was given in (2.3) in the form:

$$\lim_{x \to \infty} e^x P(N_x < \infty) = \frac{\exp\left\{-\sum_{n=1}^{\infty} n^{-1}[P_g(\ell_n \leq 0) + P(\ell_n > 0)]\right\}}{E_g(\ell_1)} .$$

Extremes in Random Fields: A Theory and its Applications, First Edition. Benjamin Yakir.
© 2013 by Higher Education Press. All rights reserved. Published 2013 by John Wiley & Sons, Ltd.

Currently we are interested in repeating the analysis using an alternative method. This alternative method will also apply a likelihood ratio identity, albeit a nonsequential one. The other components of the alternative method will bare no resemblance to the sequential methods that we used before.

In the second example we defined a scanning statistic for detecting a signal with the aid of the stochastic integral: $\int [g((x-t)/h)]^2 h^{-1} dB_x$. This scanning statistic employed a kernel g and was parameterized by the location and width of the signal $\theta = (t, h)$. Specifically, the components of the random field were the standardized score statistics that were associated with the parameter values:

$$ Z_\theta = \frac{\int g((x-t)/h)/h^{\frac{1}{2}} dB_x}{\left\{ \int [g((x-t)/h)]^2 h^{-1} dx \right\}^{\frac{1}{2}}} \, , $$

for $\theta \in T$. The goal was to approximate the probability that this two-dimensional field exceeds a high threshold z.

The analysis was one when the function g was smooth, resulting in a smooth Gaussian field, and another when the kernel was noncontinuous. In the case of a smooth kernel g we were able to use a method that is based on consideration of integral geometry in order to obtain a very accurate approximation. Keeping track only of the leading term produced a formula that was presented in (2.5):

$$ P\left(\sup_{\theta \in T} Z_\theta \geq z \right) \sim z e^{-\frac{1}{2} z^2} (2\pi)^{-3/2} \int_T |\Sigma_\theta|^{\frac{1}{2}} d\theta \, . $$

The matrix in the integral is the variance-covariance matrix of the gradient of the random field, evaluated at the parameter value.

When the Gaussian random field is not smooth the geometrical method cannot be used. Instead, the double-sum approach is an option. Specifically, for g the indicator of the interval $[-0.5, 0.5]$ we saw that the double-sum method produce the approximation:

$$ P\left(\max_{\vartheta \in T} Z_\vartheta \geq z \right) \sim z^3 e^{-\frac{1}{2} z^2} (2\pi)^{-\frac{1}{2}} \cdot (0.5)^2 (t_1 - t_0)(1/h_0 - 1/h_1) \, . $$

This approximation appeared in (2.10).

In the following sections we will give the details of the measure-transformation technique in the context of these examples. We expect, of course, to obtain the same approximations, although they may have different representations.

For convenience we divide the application of the method to several distinct steps. The details in each step may vary depending on the specifics of the application. Occasionally, an entire step may be skipped. However, we find it useful to have this conceptual organization when analyzing a new problem that is associated with the extremes of a random field.

3.2   Preliminary localization and approximation

The method that we are about to use involves a likelihood ratio identity that transforms the distribution of the random field. Consequently, the central argument is conducted in the context of that transformed distribution. However, it may be useful to start by the examination of the field in its initial distribution in order to identify the order of magnitude of the approximation of the tail and to pinpoint regions of the parameter space where extreme values are more likely to occur.

In principle, the measure transformation can be carried out directly in the context of a continuous parameter space. However, it is technically more convenient to do the analysis in a discrete setting in which some boundedness conditions are met automatically. Consequently, one may consider adding a preliminary step in which maximization over the entire parameter set is replaced by a maximization over a discrete subset. This subset should be dense enough to assure an accurate approximation of the original maximum but not too dense so as to lose the advantages of discretization. As a matter of fact, the analysis that is involved in selecting the dense subset may help in understanding the local behavior of the random field and may shed light on the characteristics of the local field that is an integral part of the main tool.

In Section 3.2.1 we carry out the preliminary localization step in the context of the first example and produce, in Section 3.2.2, a discrete approximation of the maxima in the context of the scanning statistic example.

3.2.1   Localization

A crude tool, that may be nonetheless very effective, involves the examination of the marginal tail probabilities of the random field.

Consider the first example. This example is formulated in terms of the stopping time N_x. If we wish to analyze it in the context of a random field we should first reformulate the problem as such. Indeed, the stopping time is finite if, and only if, the process (a one-dimensional field) ever crosses the threshold:

$$\{N_x < \infty\} = \left\{ \max_{1 \leq n < \infty} \ell_n \geq x \right\} .$$

The probability we seek can be obtained by the investigation of the extremes of the field ℓ_n, with $1 \leq n < \infty$ as the parameter set.

A lower bound on the probability in terms of the marginal distributions can be written as:

$$\mathrm{P}\left(\max_{1 \leq n < \infty} \ell_n \geq x \right) \geq \max_{1 \leq n < \infty} \mathrm{P}\left(\ell_n \geq x \right) .$$

The marginal probabilities depend on the value of the parameter and maximization provides information and sheds some light on the problem.

In the examination of the marginal probabilities $P(\ell_n \geq x)$ it is tempting to use the central limit theorem. Admittedly, the statistic ℓ_n is a sum of independent and identically distributed random variables. However, some caution is advised. The central limit theorem approximates the central part of the distribution. We, on the other hand, are interested in the tail of the distribution, the probability for large values of x.

In order to obtain a better approximation of the tail for a sum of independent and identically distributed random variables a likelihood ratio identity can be used:

$$P(\ell_n \geq x) = E_g(e^{-\ell_n}; \ell_n \geq x) = \int_x^\infty e^{-y} g_n(y) dy , \qquad (3.1)$$

where g_n is the density[1] of ℓ_n under the alternative distribution P_g. At this stage we are in a better position to apply the central limit theorem, with the application conducted on the distribution of ℓ_n under the alternative distribution P_g.

However, since our goal now is to approximate the density g_n, a local version of the central limit theorem, a version that estimates the density, is more appropriate. Such a local version permits the replacement of g_n, the density of ℓ_n under the alternative distribution, by the normal density with the same expectation and variance.

In the current example we consider $\ell_n = \sum_{i=1}^n \log\{g(X_i)/f(X_i)\}$, a sum of independent and identically distributed components. The expectation of each of the components is the Kullback–Leibler information index:

$$I = \int g(x) \log\{g(x)/f(x)\} dx ,$$

and the variance is $\sigma^2 = \int g(x)(\log\{g(x)/f(x)\} - I)^2 dx$.

If we substitute the density g_n by its approximation we get that

$$P(\ell_n \geq x) \approx \frac{1}{\sqrt{2\pi n\sigma^2}} \int_x^\infty e^{-y} e^{-\frac{(y-nI)^2}{2n\sigma^2}} dy = \frac{e^{-x}}{\sqrt{2\pi n\sigma^2}} \int_0^\infty e^{-z - \frac{(z+x-nI)^2}{2n\sigma^2}} dz ,$$

with the second equality following from setting $z = y - x$.

Consider the right-handside of the approximation as a function of n and x, for $x - nI = O(\sqrt{n})$ and for fixed z. The exponent in the integral is asymptotic to $-z - (x - nI)^2/(2n\sigma^2)$ and is bounded by $-z$. Applying the bounded convergence theorem we may be able to conclude that:

$$P(\ell_n \geq x) \approx \frac{e^{-x} e^{-\frac{(x-nI)^2}{2n\sigma^2}}}{\sqrt{2\pi n\sigma^2}} \int_0^\infty e^{-z} dz = \frac{e^{-x} e^{-\frac{(x-nI)^2}{2n\sigma^2}}}{\sqrt{2\pi n\sigma^2}} .$$

---

[1] We carelessly assume that ℓ_n has a density. If the distribution is discrete then integration should be replaced by a sum and densities by probabilities. Still, the formula is valid. In the general case Lebesgue–Stieltjes integration may be used.

For values of n for which In is still further away from x we get an approximation in which e^{-x} is multiplied by an exponentially vanishing function of the discrepancy between nI and x. Maximization of the marginal probability is obtained when nI gets as close as possible to x and the maximizing probability is asymptotic to $e^{-x}/\sqrt{2\pi\sigma^2 x/I}$. Observe that the exponential part of the lower bound is correct but not the polynomial part.

It is interesting to consider an upper bound which is constructed by the summation of the marginal probabilities:

$$P\left(\max_{1 \le n < \infty} \ell_n \ge x\right) = P\left(\cup_{n=1}^{\infty}\{\ell_n \ge x\}\right) \le \sum_{n=1}^{\infty} P(\ell_n \ge x).$$

If we use again the approximations of the marginal probabilities in the critical range $\{n : |In - x| = O(\sqrt{n})\} \approx \{n : |n - x/I| = O(\sqrt{x})\}$ we get that:

$$\sum_{n=1}^{\infty} P(\ell_n \ge x) \approx \frac{e^{-x}}{I} \sum_{n=x/I-C\sqrt{x}}^{x/I+C\sqrt{x}} \frac{e^{-\frac{(n-x/I)^2}{2n\sigma^2/I^2}}}{\sqrt{2\pi n\sigma^2/I^2}} \approx \frac{e^{-x}}{I},$$

provided that C is large enough. Recalling (2.3) which states that the correct approximation in this case is a constant times e^{-x} we may conclude that the upper bound produces the appropriate polynomial rate. However, the constant is still off. In summary, the crude analysis of this case proves that the correct exponential rate to use is e^{-x}. This rate is produced by sequences of length $x/I \pm O(\sqrt{x})$.

One may add rigor to the discussion by specifying a local limit theorem such as the one presented in Theorem A.1 in order to establish (3.1). However, we will take another track and apply slightly different arguments in an attempt to establish the range of parameter values in which the extremes are most likely to occur.

Consider more carefully the range of values of n that contribute to the significance level of the stopping time N_x in the first example. Fix a large C. In order to show that n larger than $n_1 = x/I + C\sqrt{x}$ need not be considered we may bound the probability that the stopping rule N_x obtains values larger than n_1:

$$P(n_1 < N_x < \infty) = E_1(e^{-\ell_{N_x}}; n_1 < N_x < \infty) \le e^{-x}P_1(n_1 < N_x).$$

Clearly, by Chebyshev's inequality,

$$P_1(n_1 < N_x) \le P_1(\ell_{n_1} < x) = P_1(\ell_{n_1} - n_1 I < x - n_1 I) \le \frac{n_1 \sigma^2}{(n_1 I - x)^2},$$

which can be made as small as one wishes by increasing C.

One may apply Kolmogorov's maximal inequality (A.2) in order to truncate the other side of the parameter space. Using the sequential likelihood ratio identity we get that:

$$P(1 \leq N_x \leq n_0) \leq e^{-x} P_1(1 \leq N_x \leq n_0) = e^{-x} P_1\left(\max_{1 \leq n \leq n_0} \ell_n \geq x\right),$$

where $n_0 = x/I - C\sqrt{x}$. From the the fact that $I > 0$ (Theorem A.12) it follows that:

$$P_1\left(\max_{1 \leq n \leq n_0} \ell_n \geq x\right) \leq P_1\left(\max_{1 \leq n \leq n_0} (\ell_n - nI) \geq x - n_0 I\right) \leq \frac{n_0 \sigma^2}{(n_0 I - x)^2},$$

which is small for large C. This completes the localization analysis for the first example.

We consider in Section 3.2.2 the second example and the approximation of a continuous parameter set by a dense sub-collection. But before doing so, let us consider the example in the context of the current topic.

In the scanning statistic example the marginal distribution of a typical element Z_θ is the standard normal distribution. This is true across the parameter set T for all values of θ. Consequently, the preliminary localization step can be skipped in this case. Still, it is useful to note that the survival function of the standard normal distribution provides a lower bound:

$$P\left(\max_{\vartheta \in T} Z_\vartheta \geq z\right) \geq \max_{\vartheta \in T} P(Z_\vartheta \geq z) = 1 - \Phi(z) \sim \frac{e^{-\frac{1}{2}z^2}}{z\sqrt{2\pi}}.$$

This lower bound produces the correct exponential term but it fails to identify the correct polynomial modification.

3.2.2   A discrete approximation

The example of scanning statistic involves a continuous parameter set T. The asymptotic expansion of the tail probability was such that

$$P\left(\sup_{\theta \in T} Z_\theta \geq z\right) \sim z^{\frac{4}{\alpha}-1} e^{-\frac{1}{2}z^2} \times \text{const.},$$

where $\alpha = 2$ corresponded to the case where the kernel is continuous, and $\alpha = 1$ corresponded to the case where the kernel was the indicator of the interval $[-0.5, 0.5]$. The constant depended on the specific selection of the kernel.

Currently, we are interested in finding a discrete subset $\hat{T} \subset T$ such that the tail probability when maximization is restricted to $\hat{T}$ is a good approximation of the tail probability of the unrestricted maxima. Specifically, we want to make sure that the choice of the subset preserves the exponential and polynomial rate. Later we will check, after the application of the measure-transformation technique, that when the subset becomes denser then the resulting constant converges to the constant that is associated with a continuous parameter set.

We produce a lower bound and an upper bound for the probability of the tail in terms of its approximation over a discrete subset. Clearly,

$$P\left(\max_{\theta \in \hat{T}} Z_\theta \geq z\right) \leq P\left(\sup_{\theta \in T} Z_\theta \geq z\right).$$

On the other hand, for any given $\epsilon > 0$,

$$P\left(\sup_{\theta \in T} Z_\theta \geq z\right) \leq P\left(\max_{\theta \in \hat{T}} Z_\theta \geq z - \frac{\epsilon}{z}\right) + P\left(\max_{\theta \in \hat{T}} Z_\theta \leq z - \frac{\epsilon}{z}, \sup_{\vartheta \in T} Z_\vartheta \geq z\right).$$

We argue that the left-hand side of the first inequality and the first of the two terms on the right-hand side of the second inequality are approximately equal to each other. Justification follows from the fact that the expansion of these terms, as a function of the threshold, produces an exponential, a polynomial, and a converging contribution. The converging contribution is the same for both terms. The ratio of the polynomial contributions converges to one since the ratio of the thresholds does. The exponential contribution is $\exp(-z^2/2)$ in the first inequality and it is $\exp\{-(z - \epsilon/z)^2/2\}$ in the second one. The ratio between the two converges to $\exp(-\epsilon)$, which can be made as close to 1 as one wishes by an appropriate selection of ϵ. Thus, in order to prove the equivalence of the discrete approximation it is sufficient to show that the error term $P(\max_{\theta \in \hat{T}} Z_\theta \leq z - \frac{\epsilon}{z}, \sup_{\vartheta \in T} Z_\vartheta \geq z)$ is of smaller order compared with the other terms that appear in the inequalities, namely $z^{\frac{4}{\alpha}-1} e^{-\frac{1}{2}z^2}$.

Associate, with each $\theta \in \hat{T}$, a local sub-region $T_\theta \subset T$ that contains θ and make sure that $T \subset \cup_{\theta \in \hat{T}} T_\theta$. The global maxima should occur in at least one of these sub-regions. Consequently, via the fact that $\{\max_{\theta \in \hat{T}} Z_\theta \leq x\} \subset \{Z_\theta \leq x\}$ and via an application of Boole's inequality:

$$P\left(\max_{\theta \in \hat{T}} Z_\theta \leq z - \frac{\epsilon}{z}, \sup_{\vartheta \in T} Z_\vartheta \geq z\right) \leq P\left(\max_{\theta \in \hat{T}} Z_\theta \leq z - \frac{\epsilon}{z}, \bigcup_{\theta \in \hat{T}} \left\{\sup_{\vartheta \in T_\theta} Z_\vartheta \geq z\right\}\right)$$

$$\leq \sum_{\theta \in \hat{T}} P\left(Z_\theta \leq z - \frac{\epsilon}{z}, \sup_{\vartheta \in T_\theta} Z_\vartheta \geq z\right).$$

The analysis proceeds by producing a bound for each of the probabilities in the sum. The bound is constructed by conditioning on the value of Z_θ and considering the supremum of the conditional field, restricted to the local sub-region.

Specifically, after dividing by the exponential and polynomial rates, we get that the re-scaled probabilities in the sum are of the form:

$$z^{-\frac{4}{\alpha}+1} e^{\frac{1}{2}z^2} P\left(Z_\theta \leq z - \frac{\epsilon}{z}, \sup_{\vartheta \in T_\theta} Z_\vartheta \geq z\right)$$

$$= z^{-\frac{4}{\alpha}+1} e^{\frac{1}{2}z^2} P\left(z\left(Z_\theta - z\right) \leq -\epsilon, \sup_{\vartheta \in T_\theta} z\left(Z_\vartheta - z\right) \geq 0\right)$$

$$= \frac{z^{-\frac{4}{\alpha}}}{\sqrt{2\pi}} \int_{-\infty}^{-\epsilon} e^{-y-\frac{y^2}{2z^2}} P\left(\sup_{\vartheta \in T_\theta} Y_\vartheta \geq 0 \mid Y_\theta = y\right) dy ,$$

that results from the fact that the distribution of $Y_\theta = z(Z_\theta - z)$ is normal with mean $-z^2$ and variance z^2.

The random field $\{Y_\vartheta : \vartheta \in T_\theta\}$, considered in the context of the conditional distribution given that $Y_\theta = y$ and restricted to the local region, is a Gaussian random field. The expectation of a component of the field is:

$$E(Y_\vartheta \mid Y_\theta = y) = y - z^2[1 - \text{Cov}(Z_\vartheta, Z_\theta)][1 + y/z^2] .$$

This expectation depends on y. The variance of a component is:

$$\text{Var}(Y_\vartheta \mid Y_\theta = y) = z^2[1 - \{\text{Cov}(Z_\vartheta, Z_\theta)\}^2]$$

and the covariance between two components is:

$$\text{Cov}(Y_\vartheta, Y_\eta \mid Y_\theta = y) = z^2[\text{Cov}(Z_\vartheta, Z_\eta) - \text{Cov}(Z_\vartheta, Z_\theta)\text{Cov}(Z_\eta, Z_\theta)] ,$$

both are independent of y. For the computation of the conditional expectation, variance, and covariance we use the properties of the multi normal distribution. (See Section A.6.)

Denote by $X_\vartheta = Y_\vartheta - E(Y_\vartheta \mid Y_\theta = y)$ the centered random field and observe that the covariance structure of the field does not depend on y. We will use the upper bound:

$$P\left(\sup_{\vartheta \in T_\theta} Y_\vartheta \geq 0 \mid Y_\theta = y\right) \leq P\left(\sup_{\vartheta \in T_\theta} X_\vartheta \geq x(y, z)\right)$$

where $x(y, z) = -y + z^2[1 - \inf_{\vartheta \in T_\theta} \text{Cov}(Z_\vartheta, Z_\theta)][1 + y/z^2]$. We will also apply a bound on the tail distribution of a Gaussian field in order to produce an upper bound for the last probability.

There are two major tools for bounding the tail of the supreme of a Gaussian field. One tool is Borell's inequality that was mentioned in the previous chapter. This inequality is based on bounding the expectation (or the median) of the supreme. An alternative tool, that does not require such bound, is Fernique's inequality. Here we will use the latter, the proof of which can be found in [8], pages 164–167.

Theorem 3.1 (Fernique's inequality). *Let $T = \prod_{j=1}^{d}[a_j, b_j]$ be a rectangle in the d-dimensional Euclidean space and let $\{X_\theta : \theta \in T\}$ be a centered Gaussian field with a bounded variance: $0 < \sigma^2 = \sup_{\theta \in T}\mathrm{Var}(X_\theta) < \infty$. Assume the $E[(X_\vartheta - X_\theta)^2] \leq \varphi(\|\vartheta - \theta\|)$, for some continuous and nondecreasing function φ that satisfies $\int_0^\infty \varphi(e^{-y^2})dy < \infty$. Then for $\lambda > 0$, $x \geq 1$, and $C > \sqrt{2d\log 2}$ we have that:*

$$P\left(\sup_{\theta \in T} X_\theta \geq x\left\{\sigma + 2(\sqrt{2} + 1)C\int_1^\infty \varphi(\sqrt{d}\lambda 2^{-y^2})dy\right\}\right)$$

$$\leq (2^d + B)e^{-\frac{1}{2}x^2}\prod_{j=1}^{d}\left(\frac{b_j - a_j}{\lambda} + \frac{1}{2}\right),$$

where $B = \sum_{n=1}^{\infty}\exp\{-2^{n-1}(C^2 - 2d\log 2)\}$.

The critical ingredient in the application of the inequality is the relation between the Euclidean distance of points in the parameter space and the expected value of the squared difference between the elements of the random field that are associated with these points. This expected squared difference is tightly linked to the covariance function since

$$E[(X_\vartheta - X_\eta)^2] = \mathrm{Var}(X_\vartheta) + \mathrm{Var}(X_\eta) - 2\mathrm{Cov}(X_\vartheta, X_\eta)$$
$$= 2z^2[1 - \mathrm{Cov}(Z_\vartheta, Z_\eta)] - z^2[\mathrm{Cov}(Z_\vartheta, Z_\theta) - \mathrm{Cov}(Z_\eta, Z_\theta)]^2.$$

We consider parameter values that are close to each other. As we discovered in the previous chapter, the expansion of the covariance function about its maximal value 1 differed between the case of a smooth Gaussian field and a Gaussian field which is not smooth.

In the smooth case we were able to compute the gradient of the covariance function and obtain that $\frac{\partial}{\partial\vartheta}\mathrm{Cov}(Z_\theta, Z_\vartheta) = \mathrm{Cov}(Z_\theta, \dot{Z}_\vartheta)$. However, in the case under consideration we also have that $\mathrm{Var}(Z_\theta) = \mathrm{Cov}(Z_\theta, Z_\theta) = 1$, for all $\theta \in T$. Consequently, when we take derivatives with respect to θ we get that:

$$0 = \frac{\partial}{\partial\theta}\mathrm{Cov}(Z_\theta, Z_\theta) = 2\mathrm{Cov}(Z_\theta, \dot{Z}_\theta) \quad \Rightarrow \quad \frac{\partial}{\partial\vartheta}\mathrm{Cov}(Z_\theta, Z_\vartheta)|_{\vartheta=\theta} = 0.$$

The conclusion is that in the smooth case $1 - \mathrm{Cov}(Z_\vartheta, Z_\theta) \leq c\|\vartheta - \theta\|^2$, for some constant c. On the other hand, in the case where the kernel was an indicator of an interval we obtained that $1 - \mathrm{Cov}(Z_\vartheta, Z_\theta) \leq c\|\vartheta - \theta\|$, for possibly a different constant c. (As a matter of fact, the relation we found was expressed in terms of the L_1 norm. However, all norms are equivalent to each other in the finite dimensional space.) One may combine the two cases using the statement:

$$1 - \mathrm{Cov}(Z_\vartheta, Z_\theta) \leq c\|\vartheta - \theta\|^\alpha,$$

with $\alpha = 2$ for the smooth case and $\alpha = 1$ for the other case.

Consider the square local sub-region $T_\theta = \theta + [0, \delta z^{-\frac{2}{\alpha}}] \times [0, \delta z^{-\frac{2}{\alpha}}]$. For any two points $\vartheta = \theta + \gamma z^{-\frac{2}{\alpha}}$ and $\eta = \theta + \xi z^{-\frac{2}{\alpha}}$ that belong to this sub-region we have that

$$E[(X_\vartheta - X_\eta)^2] \le 2c[1 + o(1)]\|\gamma - \xi\|^\alpha ,$$

with $\lim_{z \to \infty} o(1) = 0$, uniformly over the sub-region. This means that we may use the inequality in a re-parameterized setting with $\varphi(x) = c_1 x^\alpha$, with $c_1 > 2c$.

The maximization of the variance over the sub-region results in:

$$\sigma^2 = \max_{\vartheta \in T_\theta} \text{Var}(X_\vartheta) = \max_{\vartheta \in T_\theta} z^2[1 - \{\text{Cov}(Z_\vartheta, Z_\theta)\}^2] \le c_2 \delta^\alpha$$

and therefore, by taking $\lambda = \delta z^{-\frac{2}{\alpha}}$ we obtain form the application of Fernique's inequality that

$$P\left(\sup_{\vartheta \in T_\theta} X_\vartheta \ge C_\alpha x \delta^{\frac{\alpha}{2}}\right) \le B_\alpha e^{-\frac{1}{2}x^2} ,$$

for some universal constants B_α and C_α and for $x \ge 1$. For the threshold $x(y, z)$

$$x(y, z) = -y + z^2\left[1 - \inf_{\vartheta \in T_\theta} \text{Cov}(Z_\vartheta, Z_\theta)\right][1 + y/z^2] \sim -y(1 - \delta^\alpha/z^2) + d\delta^\alpha ,$$

for yet another constant d. Finally, the total number of square sub-regions T_θ that are required in order to cover T is asymptotic to $|T| \cdot \delta^{-2} z^{\frac{4}{\alpha}}$, where $|T|$ is the area of T. Putting it all together gives:

$$P\left(\max_{\theta \in \hat{T}} Z_\theta \le z - \frac{\epsilon}{z}, \sup_{\vartheta \in T} Z_\vartheta \ge z\right)$$

$$\le |T| \cdot \frac{\delta^{-2}}{\sqrt{2\pi}} \int_{-\infty}^{-\epsilon} e^{-y - \frac{y^2}{2z^2}} P\left(\sup_{\vartheta \in T_\theta} X_\vartheta \ge x(y, z)\right) dy$$

$$\le |T| \cdot \frac{B_\alpha \delta^{-2}}{\sqrt{2\pi}} \int_\epsilon^\infty e^{y - \frac{y^2}{2z^2}} e^{-\frac{1}{2}[x(y,z)]^2/[C_\alpha^2 \delta^\alpha]} dy .$$

With the divergence of z to infinity the last integrand converges to $\exp\{y - (y - d\delta^\alpha)^2/[2C_\alpha^2 \delta^\alpha]\}$. Therefore, by the dominated convergence theorem we get that the upper bound converges to

$$|T| \cdot \frac{B_\alpha \delta^{-2}}{\sqrt{2\pi}} \int_\epsilon^\infty e^{y - \frac{(y - d\delta^\alpha)^2}{2C_\alpha^2 \delta^\alpha}} dy$$

$$= |T| \cdot \frac{B_\alpha C_\alpha}{\delta^{2 - \frac{\alpha}{2}}} e^{\left(d + \frac{C_\alpha^2}{2}\right)\delta^\alpha}\left[1 - \Phi\left(\frac{\epsilon - (d + C_\alpha^2)\delta^\alpha}{C_\alpha \delta^{\frac{\alpha}{2}}}\right)\right] .$$

With the decrease in δ the upper bound, which is composed of an exponentially decreasing function of $\delta^{-\alpha}$ and a polynomially increasing function of the same quantity, is converging to 0. As a result, for each fixed $\epsilon > 0$ one may find a $\delta > 0$ so that the error that results from the approximation with a grid of span $\delta z^{-\frac{2}{\alpha}}$ is as small as one wishes.

3.3 Measure transformation

After restricting the problem to the region in the parameter space that matters and substituting a continuous parameter set by a discrete approximation one can apply the measure transformation that is at the heart of the entire approach. Frequently, this step is where art is required. The idea behind it is the old tested 'split and conquer', i.e., represent the problem of computing the probability of a maximum of the field over the set of parameters as a sum of expectations, a collection of simpler problems. Each of the elements in the sum is then approximated in the subsequent step and the approximations are summed up in the last step in order to produce the final approximation. The representation of the probability of the maximum as a sum of expectations is obtained via a likelihood ratio identity. Each parameter value that appears in the maximization is associated with a likelihood ratio and a sum of all likelihood ratios is used to produce the identity. We illustrate this step in the two examples.

Start with the first example. Take $T = \{n_0, \ldots, n_1\}$ to be the parameter set. The event of interest is:

$$A = \left\{ \max_{n \in T} \ell_n \geq x \right\} .$$

In this case, each parameter n is already associated with the likelihood $\exp(\ell_n)$, so we may as well use these likelihoods. The sum to be used in the likelihood ratio identity can for example be $\sum_{n \in T} e^{\ell_n}$, with equal weight given to each of the likelihoods.

A careful reader should be concerned by the proposal to use $\exp(\ell_n)$ as a likelihood ratio in the likelihood ratio identity applied to the event A. The likelihood is a function of only the first n observations whereas the event is a function of all observations (or, at least the observations up to n_1). That is true. However, it is worth noting that $\exp(\ell_n)$ is nonetheless a likelihood ratio for the entire collection of observations. Check that it is a likelihood ratio for the alternative distribution of independent observations that assign the density g to the first n observations and the density f to the ensuing ones. Denote this alternative distribution by P_n (and the resulting expectation by E_n).

A more general justification to the statement that $\exp(\ell_n)$ is a likelihood ratio follows from the fact that this random variable is non-negative and its expectation, where the expectation is taken with respect to entire collection of observations and under the null hypothesis, is equal to one. Such random variables are likelihood ratios for an alternative distribution with a density that is produced by the multiplication of the null joint density of the observations times the given random variable.

We are now in the position to apply the likelihood ratio identity:

$$P(A) = E\left(\frac{\sum_{n\in T}e^{\ell_n}}{\sum_{m\in T}e^{\ell_m}}; A\right) = \sum_{n\in T}E_n\left(\frac{1}{\sum_{m\in T}e^{\ell_m}}; A\right),$$

where the last equality follows from changing the order of summation and expectation and the application of the likelihood ratio identity to each of the summands.

We rearrange each term and add notations:

$$E_n\left(\frac{1}{\sum_{m\in T}e^{\ell_m}}; A\right) = e^{-x}E_n\left(\frac{\max_{m\in T}e^{\ell_m-\ell_n}}{\sum_{m\in T}e^{\ell_m-\ell_n}}e^{-[\ell_n-x+\max_{m\in T}(\ell_n-\ell_n)]}; A\right)$$

$$= e^{-x}E_n\left(\frac{M_n}{S_n}e^{-[\tilde\ell_n+m_n]}; \tilde\ell_n+m_n \geq 0\right),$$

where $S_n = \sum_{m\in T}e^{\ell_m-\ell_n}$ is the sum of likelihood ratios, $M_n = \max_{m\in T}e^{\ell_m-\ell_n}$ is the maximal likelihood ratio, $m_n = \log M_n$, and $\tilde\ell_n = \ell_n - x$. In the last equality we rephrased the event A in terms of the newly defined random variables:

$$A = \left\{\max_{n\in T}\ell_n \geq x\right\} = \{\tilde\ell_n+m_n \geq 0\}.$$

The essence of the final representation of a summand is the dissection to a large deviations exponential decay, given by the exponent e^{-x}, and lower order contributions that reside in the expectation. The random variables in expectation are further dissected into random variables that are influenced mainly by local perturbations and the random variable that captures the main part of the variability. The latter is the random variable $\tilde\ell_n$, which has expectation $nI - x$ under the alternative and variance $n\sigma^2$. We call this random variable *the global term*. The other random variables are S_n, M_n (and its log). These random variables are functions of what we call *the local field*, with elements, that in the current case are of the form $\ell_m - \ell_n$ and are parameterized by m.

The localization theorem of the next section investigates the limiting joint distribution of the global term and the local field. If the situation is that the two components are asymptotically independent then the expectation can be dissected as a product of two terms. One is the expectation of the limit of the ratio between M_n and S_n and the other is similar to the expansion of the marginal probability $P(\ell_n \geq x)$ that was discussed informally in the previous section. We leave the details to the next section and to Chapter 5 that is devoted to the localization theorem.

Consider next the second example of a Gaussian random field. In this example the elements of the field are the statistics Z_θ, for $\theta \in T$. Unlike the other example, these statistics are not log-likelihood ratios, since their exponent does not integrate to 1. Therefore, we cannot use them directly for the measure transformation as we did before. However, we can still produce for each Z_θ a likelihood ratio, for example by means of exponential tilting.

In general, exponential tilting for a random variable X with density (or probability mass function) dP corresponds to the alternative distribution which is specified by a likelihood ratio $\exp\{\xi X - \psi(\xi)\}$. The normalizing term, the log-moment generating function, is defined via $\psi(\xi) = \log \mathrm{E} \exp\{\xi X\}$. Clearly, the density of the resulting alternative distribution is $\mathrm{dP}_\xi = \exp\{\xi X - \psi(\xi)\}\mathrm{dP}$ and it is defined for each ξ that results in a finite log-moment generating function. The user is free to choose the ξ that is most convenient for a given application.

Exponential tilting, when the distribution of Z_θ is standard normal, refers to the likelihood ratio $\exp(\mu Z_\theta - \mu^2/2) = \exp(\ell_\theta)$, where μ is a number we may specify for our convenience. The alternative distribution of Z_θ after tilting is the normal distribution with expectation μ and with variance 1. Denote this distribution by P_θ. Just as before, we should consider this distribution not only in the context of the specific Z_θ but in the context of the joint distribution of the entire field. Namely, we should ask ourselves what is the alternative joint distribution of the field that produces ℓ_θ as a log-likelihood ratio?

In order to answer this question consider a (finite) collection of normal random variables $Z = \{Z_\vartheta\}$ with zero mean and unit variance, written as a column vector. Let Σ be the variance-covariance matrix of this collection. Consider an alternative distribution for this collection with the same variance-covariance matrix but with an expectation vector η. The resulting log-likelihood ratio is $\eta'\Sigma^{-1}Z - (\eta'\Sigma^{-1}\eta)/2$. We can obtain the form of a tilted log-likelihood ratio if we take $\eta = \mu\rho$, where ρ is the column of the matrix Σ that is associated with the variable $Z_\theta \in \{Z_\vartheta\}$. This is the case since $\eta'\Sigma^{-1}Z = \mu Z_\theta$ and $(\eta'\Sigma^{-1}\eta)/2 = \mu^2/2$. To conclude, the alternative distribution P_θ assigns a Gaussian distribution to the field. The covariance structure under the alternative distribution is identical to the covariance structure under the null. The expectation is shifted. Under the alternative distribution the expectation of the component Z_ϑ is equal to $\mu \cdot \mathrm{Cov}(Z_\vartheta, Z_\theta)$, for $\vartheta \in T$.

We have still the freedom to select the numerical value of μ. We choose the value that will produce the correct large deviations exponential decay for the marginal probability $\mathrm{P}(Z_\theta \geq z)$. The resulting value is $\mu = z$, which produces the log-likelihood ratio $\ell_\theta = zZ_\theta - z^2/2$.

Misusing the notations somewhat we denote by T the discrete approximation of the original continuous parameter set. In the examples that we use the discrete parameter set is of the form

$$T = \left\{(t, h) : t = i\delta z^{-1}, h = j\delta z^{-1}, t_0 \leq t \leq t_1, h_0 \leq h \leq h_1\right\}$$

in the case of a continuous kernel and it is of the form:

$$T = \left\{(\theta_1, \theta_2) : \theta_1 = i\delta z^{-2}, \theta_2 = j\delta z^{-2}, t_0 \leq (\theta_1 + \theta_2)/2 \leq t_1, h_0 \leq \theta_2 - \theta_1 \leq h_1\right\}$$

when the kernel is an indicator of an interval.

In either case, let A be the event of interest:

$$A = \left\{ \max_{\theta \in T} Z_\theta \geq z \right\} .$$

In the zero-mean unit-variance Gaussian case with the discrete approximation of the parameter space we use the sum of the likelihood ratios $\sum_{\theta \in T} e^{\ell_\theta}$ for the measure transformation:

$$P(A) = E\left(\frac{\sum_{\theta \in T} e^{\ell_\theta}}{\sum_{\vartheta \in T} e^{\ell_\vartheta}}; A \right) = \sum_{\theta \in T} E_\theta \left(\frac{1}{\sum_{\vartheta \in T} e^{\ell_\vartheta}}; A \right) .$$

Each element in the sum may be represented more conveniently as:

$$E_\theta \left(\frac{1}{\sum_{\vartheta \in T} e^{\ell_\vartheta}}; A \right)$$

$$= e^{-\frac{1}{2}z^2} E_\theta \left(\frac{\max_{\vartheta \in T} e^{\ell_\vartheta - \ell_\theta}}{\sum_{\vartheta \in T} e^{\ell_\vartheta - \ell_\theta}} e^{-[\ell_\theta - \frac{1}{2}z^2 + \max_{\vartheta \in T}(\ell_\vartheta - \ell_\theta)]}; A \right)$$

$$= e^{-\frac{1}{2}z^2} E_\theta \left(\frac{\max_{\vartheta \in T} e^{z(Z_\vartheta - Z_\theta)}}{\sum_{\vartheta \in T} e^{z(Z_\vartheta - Z_\theta)}} e^{-[z(Z_\theta - z) + \max_{\vartheta \in T} z(Z_\vartheta - Z_\theta)]}; A \right)$$

$$= e^{-\frac{1}{2}z^2} E_\theta \left(\frac{M_\theta}{S_\theta} e^{-[\tilde{\ell}_\theta + m_\theta]}; \tilde{\ell}_\theta + m_\theta \geq 0 \right) ,$$

where $S_\theta = \sum_{\vartheta \in T} e^{z(Z_\vartheta - Z_\theta)}$ replaces S_n, $M_\theta = \max_{\vartheta \in T} e^{z(Z_\vartheta - Z_\theta)}$ replaces M_n, $m_\theta = \log M_\theta$, and $\tilde{\ell}_\theta = z(Z_\theta - z)$. Again, we rephrased the event A in terms of the newly defined random variables:

$$A = \left\{ \max_{\theta \in T} Z_\theta \geq z \right\} = \{\tilde{\ell}_\theta + m_\theta \geq 0\} .$$

It is worth noting that while the large deviations decay in the Gaussian example is not the same as the one obtained for the example of sequential testing, still the structure of the expectation that captures lower-order contributions is the same. In the second problem, as in the first, that expectation involves a global term and a local field. The global term in the Gaussian case is $z(Z_\theta - z)$ and the local field is $z(Z_\vartheta - Z_\theta)$, parameterized by ϑ.

The distinction between the case of a smooth Gaussian field and the case where the field is continuous but not smooth is expressed in the specifications of the grid that is used in order to approximate the continuous parameter set. In the situation of a non-smooth field a denser grid is required. However, once the grid T is specified, the resulting representation of the tail probability via measure transformation is identical for both cases. Differences will emerge again when we investigate in the next section the convergence of the components in the sum and, of course, in the finite step of obtaining the overall approximation, which is the subject of the last section.

3.4   Application of the localization theorem

This step is usually the most technically involved part of the proof. It deals with the limit of terms of the form:

$$E[(M_\kappa/S_\kappa)\exp[-(\tilde{\ell}_\kappa + \log M_\kappa)]; \tilde{\ell}_\kappa + \log M_\kappa \geq 0],$$

where κ is associated with n in the example of sequential testing and κ is associated mainly with z in the example of a scanning statistic, but also with θ. These expectations emerged as elements in the representation of the tail probability of the maxima of the field. The expectation corresponds to an appropriate alternative distribution, not the original null distribution. The representation is a function of a 'global' term $\tilde{\ell}_\kappa$ and a 'local' random field. The former is associated with the specific log-likelihood ratio and the latter corresponds to the difference between other log-likelihood ratios and the specific one. The approximation identifies the asymptotic distribution of the local field and the asymptotic independence between it and the global term. The limit is given in terms of a bounded functional of the limit of the local field and in terms of the density of the global element, evaluated at the origin.

In the case where the parameter set is continuous or very refined we may choose to approximate the maximization of the random field by the maximization over a cruder subset of the parameter space. The approximating subset should be dense enough to assure accurate approximation. That was illustrated above in the context of the scanning statistic. On the other hand, the subset should not be too dense. The analysis of the local field sheds light on the last statement. The limit of the local field should be such that the resulting functional is not only bounded but also strictly positive.

The localization theorem is stated in Chapter 5 in terms of an abstract index κ. Convergence to a limit occurs as $\kappa \to \infty$. In the context of a specific application the value of the parameter can be identified by the examination of the distribution of the global term $\tilde{\ell}_\kappa$. In the formulation of the localization theorem, $\kappa^{\frac{1}{2}}$ is the multiplicative factor that balances the rate of convergence of the density of $\tilde{\ell}_\kappa$ to 0. Consequently, one may typically equate κ with the variance of the global term or choose it to be asymptotically proportional to that variance.

The statement of the localization theorem involves also a local σ-algebra denoted by $\hat{\mathcal{F}}_\kappa$. This local σ-algebra is asymptotically independent of $\tilde{\ell}_\kappa$ but it carries enough information to construct approximations of the random variables that summarize the contributions of the local field. These random variables, denoted $\hat{M}_\kappa$ and $\hat{S}_\kappa$, are measurable with respect to $\hat{\mathcal{F}}_\kappa$ and serve as approximations of the original random variables M_κ and S_κ, respectively. The expectation of the ratio of these approximating random variables converges to a limit that is used in the formulation of the approximation of the tail probability.

Specifically, in the example of sequential testing we may take $\kappa = n$. One may then choose some finite τ and use $\hat{\mathcal{F}}_\kappa = \sigma\{X_{n-\tau}, \ldots, X_{n+\tau}\}$, for the observations in the vicinity of n. The local field $\{\ell_m - \ell_n\}$, restricted to the region

$|m - n| \leq \tau$, is measurable with respect to the given σ-algebra. Producing $\hat{M}_\kappa$ and $\hat{S}_\kappa$ by restricting the maximization (respectively, summation) to the reduced region of parameter values will result in random variables that approximate the original maximization (or summation), as long as τ is large enough. The distribution of the ratio $\hat{M}_\kappa / \hat{S}_\kappa$ is independent of κ, for all κ not too close to n_0 or n_1. Hence, the expectation of the ratio is unchanged in the limit. This expectation approximates a similar expectation that involves a ratio in which $\tau = \infty$, defined for a double-ended sequence of observations that extends for all m, $-\infty < m < \infty$. That last functional is the term that is denoted by $E(\mathcal{M}/\mathcal{S})$ and will be used in the approximation of the significance level of the sequential test.

For the example of a scanning statistic we may use $\kappa = z^2$, but keep track of the value of θ. The local σ-algebra involves the local field $z(Z_\vartheta - Z_\theta)$ and can be taken to be $\sigma\{z(Z_\vartheta - Z_\theta) : \|\vartheta - \theta\| \leq \tau z^{-\frac{2}{\alpha}}\}$, for some large τ. If the field is smooth we use $\alpha = 2$. For the non continuous kernel we use $\alpha = 1$. As a consequence of the approximation of the continuous parameter set by a discrete one we obtain that the number of parameter values in $\{\vartheta : \|\vartheta - \theta\| \leq \tau z^{-\frac{2}{\alpha}}\}$ is finite. Again, we produce $\hat{M}_\kappa$ and $\hat{S}_\kappa$ by restricting maximization and summation to the smaller subset of parameter values. In the limit, as $\kappa \to \infty$, we get the the local random field $z(Z_\vartheta - Z_\theta)$ converges to a limit Gaussian field. Subsequently, the functional $E(\hat{M}_\kappa / \hat{S}_\kappa)$ converges to a limit $E[\hat{\mathcal{M}}/\hat{\mathcal{S}}]$ that can be expressed in terms of the limit Gaussian field, restricted to the subset of parameter values. The term that appears as the outcome of the limit in the statement of the localization theorem, namely $E(\mathcal{M}/\mathcal{S})$, is obtained by letting τ go to infinity.

The localization theorem is formulated as a collection of conditions and a statement of a limit. Given an application, if the conditions can be verified then the limit is validated. In Chapter 5 we state and prove two versions of the theorem. For pedagogical reasons we give first a relatively simple formulation. Later, we present a more complex formulation with conditions that are usually easier to verify. In the current section we verify the conditions for the two examples that we are tracking. Both examples are simple enough and can be validated using the simpler formulation of the theorem. However, in order to practice the more useful formulation we carry out the validation using the more complex formulation.

Specifically, we need to check five conditions. Given $\epsilon, \epsilon_3 > 0$, for a function $g(\kappa)$ to be specified for the given application but obeying the relation $\log \kappa \leq g(\kappa) \leq \epsilon \kappa^{\frac{1}{2}}$, for some $C < \infty$, and for all large κ we should have:

I*: M_κ, S_κ, $\hat{M}_\kappa$ and $\hat{S}_\kappa$ satisfy $0 \leq M_\kappa / S_\kappa \leq C$ and $0 \leq \hat{M}_\kappa / \hat{S}_\kappa \leq C$ with probability one.

II*: Denote $A_{II}^c = \{|\log M_\kappa - \log \hat{M}_\kappa| > \epsilon\} \cup \{|\hat{S}_\kappa / S_\kappa - 1| > \epsilon\}$. For some $0 < \delta$ that does not depend on ϵ:

$$\max_{|x| \leq 3g(\kappa)} P(A_{II}^c \cap \{\tilde{\ell}_\kappa + \hat{m}_\kappa \in x + (0, \delta]\} \cap \{|\hat{m}_\kappa| \leq g(\kappa)\}) \leq \epsilon \kappa^{-\frac{1}{2}},$$

where $\hat{m}_\kappa = \log \hat{M}_\kappa$.

III*: $E[\hat{M}_\kappa / \hat{S}_\kappa]$ converges to $E[\hat{\mathcal{M}}/\hat{\mathcal{S}}]$ and $|E[\hat{\mathcal{M}}/\hat{\mathcal{S}}] - E[\mathcal{M}/\mathcal{S}]| \leq \epsilon_3$.

IV*: There exist $\mu \in \mathbb{R}$ and $\sigma \in \mathbb{R}^+$ such that for every $0 < \epsilon_4, \delta$, for any event $E \in \hat{\mathcal{F}}_\kappa$ having boundary measure 0, and for all large enough κ:

$$\sup_{|x| \leq 3g(\kappa)} |\kappa^{\frac{1}{2}} P(\tilde{\ell}_\kappa + \hat{m}_\kappa \in x + (0, \delta], E) - \frac{\delta}{\sigma} \phi\left(\frac{\mu}{\sigma}\right) P(E)| \leq \epsilon_4,$$

and also:

V*: $P(|\log M_\kappa| > g(\kappa))$, $P(|\log \hat{M}_\kappa| > g(\kappa))$ and $P(\log M_\kappa - \log \hat{M}_\kappa < -\epsilon)$ are all $o(\kappa^{-\frac{1}{2}})$.

Theorem 5.2 states that if Conditions I*–V* hold then:

$$\lim_{\kappa \to \infty} \kappa^{\frac{1}{2}} E\left[(M_\kappa / S_\kappa) e^{-(\tilde{\ell}_\kappa + \log M_\kappa)}; \tilde{\ell}_\kappa + \log M_\kappa \geq 0\right] = \sigma^{-1} \phi(\mu/\sigma) E[\mathcal{M}/\mathcal{S}].$$

We now consider the list of conditions, not in the above order, and check them for the example of sequential testing and for the example of a scanning statistic.

3.4.1   Checking Condition I*

Condition I* is straightforward. In both cases we have that the ratio is between a maximum of likelihood ratios – non-negative statistics – and the sum of the same likelihood ratios. Since the sum of non-negative terms is always larger than the largest term we get that the ratio is bounded by $C = 1$. This is the case when the collection of likelihood ratios is unrestricted as well as when it is restricted to belong to the local sub-collection of parameter values.

3.4.2   Checking Condition V*

Condition V* involves three probabilities. The last probability corresponds to the event where the maximization of the likelihood ratios over the entire parameter set is less than the maximization over the local sub-collection of parameter values. Clearly, the former is larger. Therefore, the probability of the given event is 0 and that part of the condition is trivially met. Using the same argument, we may claim that the first of the two remaining probabilities is always larger than the second probability. Consequently, it is sufficient to show that the first probability meets the condition. Moreover, in both examples the trivial likelihood ratio 1 is among the likelihood ratios that participate in the maximization. Consequently, $\hat{M}_\kappa \geq 1$ and we need not worry about negative values of the log of this statistic.

The first probability, in the case of sequential testing, may be written as:

$$P_n(M_n > e^{g(n)}) = P_n\left(\max_{n_0 \leq m \leq n_1} e^{\ell_m - \ell_n} > e^{g(n)}\right) \leq (n_1 - n_0) e^{-g(n)}.$$

The last inequality follows from an application of Boole's inequality, together with Markov's inequality. The expectation $E_n(e^{\ell_m - \ell_n})$ is equal to 1 since the random variable $e^{\ell_m - \ell_n}$ is a likelihood ratio. All the values in the interval $[n_0, n_1]$ are asymptotic to x/I and the number of values in the interval is proportional to $x^{\frac{1}{2}}$. As a result, the function $g(n) = (1 + \epsilon) \log(n)$ will do the job.

For the case of a scanning statistic we have that the number of parameters in T is asymptotically proportional to $z^{4/\alpha}$. Arguing as before we get that:

$$P_\theta(M_\theta > e^{g(z^2)}) = P_\theta\left(\max_{\vartheta \in T} e^{z(Z_\vartheta - Z_\theta)} > e^{g(z^2)}\right) \leq |T| e^{-g(z^2)},$$

where $|T|$ is the cardinality of the set T. Taking $g(z^2) = (4/\alpha + 1 + \epsilon) \log z$ is sufficient.

3.4.3   Checking Condition IV*

Turn to Condition IV*, first for the scanning statistic then for sequential testing. In the former case the joint distribution of global term $z(Z_\theta - z)$ and the local field $\{z(Z_\theta - Z_\theta)\}$ is multivariate normal. The expectation of the global term under the alternative distribution is $z E_\theta(Z_\theta - z) = 0$ and the variance is $z^2 \text{Var}_\theta(Z_\theta) = z^2$. It follows that $\mu = 0$ and $\sigma^2 = 1$. The density of the global term can be approximated by $(2\pi z^2)^{-\frac{1}{2}}$ uniformly in the range $|x| \leq 3(4/\alpha + 1 + \epsilon) \log z$. In order to verify Condition IV* it is sufficient to show that the conditional covariance structure of the local random field, given the global term, converges to the unconditioned covariance structure and that the conditional expectations of the local field converge to the unconditional ones, uniformly in the given range of values of the global term. However,

$$z^2[\text{Cov}(Z_\vartheta - Z_\theta, Z_\eta - Z_\theta | Z_\theta) - \text{Cov}(Z_\vartheta - Z_\theta, Z_\eta - Z_\theta)]$$
$$= -z^2[1 - \text{Cov}(Z_\vartheta, Z_\theta)][1 - \text{Cov}(Z_\eta, Z_\theta)],$$

which converges to zero, when $z \to \infty$, for all ϑ and η in the local region. Likewise,

$$z\{E_\theta[Z_\vartheta - Z_\theta | z(Z_\theta - z) = x] - E_\theta[Z_\vartheta - Z_\theta]\} = z^2[1 - \text{Cov}(Z_\vartheta, Z_\theta)]x/z^2,$$

converges to zero, uniformly in x, in the restricted range of x values. This completes the validation of the condition for scanning statistics.

Consider sequential testing. In this case the expectation of the global term is $E_n(\ell_n - x) = nI - x$, for $I = \int g(x) \log\{f(x)/g(x)\}dx$, and the variance is $n\sigma^2$, for $\sigma^2 = \int g(x)(\log\{f(x)/g(x)\} - I)^2 dx$. Since $\kappa = n$ we obtain that σ^2 is the value that is used in the declaration of Condition IV* and

$$\mu = \lim_{\kappa \to \infty} \frac{E_\kappa(\tilde{\ell}_k)}{\kappa^{\frac{1}{2}}} = \lim_{x \to \infty} \frac{In - x}{n^{\frac{1}{2}}},$$

with values that range between $-I^{\frac{3}{2}}C$ and $I^{\frac{3}{2}}C$.

The local σ-algebra $\hat{\mathcal{F}}_n = \sigma\{X_{n-\tau}, \ldots, X_{n+\tau}\}$ is generated by observations in the range $[n - \tau, n + \tau]$. One may write the global term as a sum of two independent random variables: $\ell_n - x = (\ell_{n-\tau-1} - x) + (\ell_n - \ell_{n-\tau-1})$, the first of which is independent of the local σ-algebra and the second is measurable with respect to it. We may establish Condition IV* by conditioning on the local σ-algebra and applying a local limit theorem such as Theorem A.1 to the independent component $\ell_{n-\tau-1} - x$:

$$P_n(\tilde{\ell}_\kappa \in y + (0, \delta], E) = E_n[P_n(\ell_{n-\tau-1} - x \in y - (\ell_n - \ell_{n-\tau-1}) + (0, \delta]|\hat{\mathcal{F}}_n); E],$$

where $y = \xi - \hat{m}_n$, for any ξ, $|\xi| \leq 3(1 + \epsilon)\log n$. The asymptotic expectation and variance of $\ell_{n-\tau-1} - x$ are the same as those of $\ell_{n-1} - x$. If the distribution of components is non lattice then the normal density emerges as the limit, uniformly in $|y| \leq 4(1 + \epsilon)\log n$ and over the event $\{|\ell_n - \ell_{n-\tau-1}| < \epsilon n^{\frac{1}{2}}\} \cap \{\hat{m}_\kappa \leq (1 + \epsilon)\log n\}$. Proving that the probability of the complementary of this event is $o(n^{-\frac{1}{2}})$ completes the proof. However, for the complementary of the event for $\hat{m}_n$ we have Condition V*. For the complementary of the other event we get from Markov's inequality that:

$$P_n\left(|\ell_n - \ell_{n-\tau-1}| \geq \epsilon n^{\frac{1}{2}}\right) \leq \frac{\tau\sigma^2}{(\epsilon n^{\frac{1}{2}} - \tau I)^2}$$

as required.

3.4.4 Checking Condition II*

Condition II* is more demanding. We analyze the condition by the identification of an event that contains A_{II}^c and produce a bound for the event. For both cases we use the fact that $\hat{M}_\kappa \leq M_\kappa$, $\hat{S}_\kappa \leq S_\kappa$. Consequently, one of the directions that is implied by the use of absolute values in the definition of A_{II}^c is obtained for free. Still we need to bound the probability of the event

$$A_{\text{II}}^c = \{M_\kappa > \hat{M}_\kappa e^\epsilon\} \cup \{\hat{S}_\kappa < (1 - \epsilon)S_\kappa\},$$

intersected with an event formulated in terms of the global term and the log of the local maxima. From the fact that $1 \leq \hat{M}_\kappa$, $1 \leq \hat{S}_\kappa$, and $M_\kappa \leq \hat{M}_\kappa + S_\kappa - \hat{S}_\kappa$ we get that:

$$\{M_\kappa > \hat{M}_\kappa e^\epsilon\} \cup \{\hat{S}_\kappa < (1 - \epsilon)S_\kappa\} \subset \{S_\kappa - \hat{S}_\kappa > \epsilon\}$$

since $\min\{e^\epsilon - 1, \epsilon/(1 - \epsilon)\} \geq \epsilon$.

For both cases, the statistic $S_\kappa - \hat{S}_\kappa$ is a sum of likelihood ratios. In the case of sequential testing $S_\kappa - \hat{S}_\kappa = \sum_{\{m:|m-n|>\tau\}} \exp\{\ell_m - \ell_n\}$ and in the case of the scanning statistic $S_\kappa - \hat{S}_\kappa = \sum_{\{\vartheta:\|\vartheta-\theta\|>\tau z^{-\frac{2}{\alpha}}\}} \exp\{z(Z_\vartheta - Z_\theta)\}$. Associate with each parameter value outside the local region a non-negative quantity (p_m

in sequential testing and p_ϑ is the scanning statistic problem) such that the sum of quantities is bounded by 1. It follows that:

$$A_{II}^c \subset \bigcup_{\{m:|m-n|>\tau\}} \{\ell_m - \ell_n \geq \log(\epsilon p_m)\}$$

in the case of sequential testing and:

$$A_{II}^c \subset \bigcup_{\{\vartheta:\|\vartheta-\theta\|>\tau z^{-\frac{2}{\alpha}}\}} \{z(Z_\vartheta - Z_\theta) \geq \log(\epsilon p_\vartheta)\}$$

in the case of scanning for a signal hidden in Gaussian white noise. The verification of Condition II* proceeds by the investigation of the probability of the events in the unions when they are intersected with the event $\{\ell_n - x \in y + (0, \delta]\}$, $y = \xi - \hat{m}_n$, in the first case and the event $\{z(Z_\theta - z) + \hat{m}_\theta \in x + (0, \delta]\}$ in the second case.

Consider the problem of sequential testing. Up until this point we were able to get away with making the minimal assumption that the distribution of the increments of the log-likelihood ratio under the alternative distribution are non lattice and have a finite second moment. For the current condition this will not be enough. Specifically, at this point we would like to employ the Berry–Esseen theorem in order to produce a uniform bound in the local limit approximation. This theorem requires the existence of a third moment. This is probably sufficient. However, assuming a fourth moment, which implies the existence of a third moment, simplifies the proof further. Thus, we make that assumption: $\int g(x)[\log\{f(x)/g(x)\} - I]^4 dx = \mu_4 < \infty$.

The alternative expectation of the log-likelihood statistics $\ell_m - \ell_n$ that form the local field is equal to $-I|m - n|$. Let $U_m - U_n = \ell_m - \ell_n + |m - n| \cdot I$ be the centered sum of increments and take $p_m = 0.5 \exp\{\tau/2\} \exp\{-|m - n|(I/2)\}$.

Look at the case $m > n + \tau$. Intersecting with the event $\{\ell_n - x \in y + (0, \delta]\}$, that is conditionally independent of $U_m - U_n$, given the local σ-algebra $\hat{\mathcal{F}}_n$, we get that:

$$P_n \left(\{\ell_m - \ell_n \geq \log(\epsilon p_m)\} \cap \{\ell_n - x \in y + (0, \delta]\}\right)$$

$$= E\left[P_n\left(U_m - U_n \geq \log(\epsilon p_m) + (m - n)I \mid \hat{\mathcal{F}}_n\right) P_n(\ell_n - x \in y + (0, \delta]|\hat{\mathcal{F}}_n)\right]$$

$$\leq \frac{(m - n)\mu_4 + 3(m - n)^2\sigma^4}{\{\log(0.5\epsilon) + \tau/2 + (m - n)(I/2)\}^4} \times cn^{-\frac{1}{2}},$$

for some finite constant c. The last inequality is a result of the uniform local limit produced by the Berry–Esseen theorem and applied to the distribution of $\ell_{n-\tau}$ that is independent of the local σ-algebra. The bound on the factor on the left-hand side is produced by the application of the Markov inequality to $(U_m - U_n)^4$. The expectation of this random variable is bounded by the numerator of the ratio on the left-hand side.

In the case where $n_0 \le m < n - \tau$ we may not use independence directly. However, essentially the same proof still works. Denote $A_m = \{U_m - U_n \ge \log(\epsilon p_m) + (m-n)I\}$. Condition on the value of $U_n - U_m$ and on the local σ-algebra to obtain:

$$P_n \left(\{\ell_m - \ell_n \ge \log(\epsilon p_m)\} \cap \{\ell_n - x \in y + (0, \delta]\} \right)$$
$$= E_n \left[P_n \left(U_n \in x - nI + y + (0, \delta] \mid U_m - U_n, \hat{\mathcal{F}}_n \right); A_m \right]$$
$$= E_n \left[P_n \left(U_m \in x - nI + y + U_m - U_n + (0, \delta] \mid U_m - U_n, \hat{\mathcal{F}}_n \right); A_m \right]$$

Currently, U_m is independent of $U_n - U_m$ and the local σ-algebra. Applying the Berry–Esseen theorem we get that the conditional probability is bounded by $cm^{-\frac{1}{2}}$, which produces the inequality:

$$\le \frac{(m-n)\mu_4 + 3(m-n)^2 \sigma^4}{\{\log(0.5\epsilon) + \tau/2 + (m-n)(I/2)\}^4} \times (n/m)^{\frac{1}{2}} \times cn^{-\frac{1}{2}} .$$

Summing over all parameters in the range and approximating the sum by an integral produces the asymptotic inequality:

$$n^{\frac{1}{2}} P_n (A_{II}^c \cap \{\ell_n - x \in y + (0, \delta]\})$$
$$\underset{\sim}{\le} \frac{1}{\tau} \int_1^\infty \frac{2c[1 + 2C/(Ix^{\frac{1}{2}})][y\mu_4/\tau + 3y^2\sigma^2]}{[\log(0.5\epsilon)/\tau + 0.5 + y(I/2)]^4} dy .$$

The right-hand side is $O(1/\tau)$. Choosing τ large enough finishes the validation of Condition II^*.

Consider the same condition in the Gaussian setting. Here we may compute the probabilities involved in the bound by conditioning on the value the global term $z(Z_\theta - z)$ and using the fact that the density of the global term is bonded by its value at the origin, namely: $P_\theta(z(Z_\theta - z) = x) \le [2\pi z^2]^{-\frac{1}{2}}$. However, we do intend to treat the bound on the values that $\hat{m}_\theta$ may obtain more carefully than before.

Given a value of x, split the probability between the event $\{\hat{m}_\theta \le m\}$ and its complementary to get:

$$P_\theta(A_{II}^c \cap \{z(Z_\theta - z) + \hat{m}_\theta \in x + (0, \delta]\})$$
$$\le P_\theta(A_{II}^c \cap \{-m < X_\theta \le \delta\}) + E_\theta[P_\theta(\{X_\theta + \hat{m}_\theta \in (0, \delta]\} | \hat{\mathcal{F}}_\theta); \hat{m}_\theta > m]$$

where $X_\theta = z(Z_\theta - z) - x$. The first probability is bounded by:

$$\frac{m + \delta}{z\sqrt{2\pi}} \times \max_{x-m < y \le x+\delta} P(A_{II}^c | z(Z_\theta - z) = y)$$

and the second term is dominated by:

$$\left\{ 1 - \sum_{\{\vartheta : \|\vartheta - \theta\| \le \tau z^{-\frac{2}{\alpha}}\}} [1 - \text{Cov}(Z_\theta, Z_\vartheta)]^2 \right\}^{-\frac{1}{2}} \frac{1}{z\sqrt{2\pi}} P_\theta(\hat{m}_\theta > m) .$$

The term in the curly brackets is the conditional variance of Z_θ, given the local σ-algebra $\hat{\mathcal{F}}_\theta$, and is converging to 1 with the increase in the threshold. By Condition III* we obtain that the random variable $\hat{m}_\theta$ converges to a finite random variable. Consequently, we can make the product of z and the second term as small as we wish by the selection of a large enough m. The overall proof of Condition II* will be complete once we produce a uniform bound for the probabilities $P(A_{\text{II}}^c | z(Z_\theta - z) = y)$.

The conditional probability for an event in the union that contains A_{II}^c is a function of the conditional expectation and conditional variance of the increment of the local field that defines the event. The conditional expectation of the local increment, given the global term, is:

$$E_\theta[z(Z_\vartheta - Z_\theta) | z(Z_\theta - z) = y] = -z^2[1 - \text{Cov}(Z_\vartheta, Z_\theta)][1 + y/z^2] ,$$

and the conditional variance, independent of y, is:

$$\text{Var}_\theta[z(Z_\vartheta - Z_\theta) | z(Z_\theta - z)]$$
$$= z^2[1 - \{\text{Cov}(Z_\vartheta, Z_\theta)\}^2] < 2z^2[1 - \text{Cov}(Z_\vartheta, Z_\theta)] .$$

We may bound the conditional probability of a typical event in the union that contains A_{II}^c by bounding the conditional expectation from below and bounding the conditional variance from above:

$$P_\theta(z(Z_\vartheta - Z_\theta) \ge \log(\epsilon p_\vartheta) | z(Z_\theta - z) = y)$$
$$\le 1 - \Phi\left(\frac{\log(\epsilon p_\vartheta) + z^2[1 - \text{Cov}(Z_\vartheta, Z_\theta)][1 - \epsilon]}{\sqrt{2}z[1 - \text{Cov}(Z_\vartheta, Z_\theta)]^{\frac{1}{2}}} \right) ,$$

for all $y > -\epsilon z^2$. For each y in the range:

$$P_\theta(A_{\text{II}}^c | \{z(Z_\theta - z) = y)$$
$$\le \sum_{\{\vartheta : \|\vartheta - \theta\| > \tau z^{-\frac{2}{\alpha}}\}} \left\{ 1 - \Phi\left(\frac{\log(\epsilon p_\vartheta) + z^2[1 - \text{Cov}(Z_\vartheta, Z_\theta)][1 - \epsilon]}{\sqrt{2}z[1 - \text{Cov}(Z_\vartheta, Z_\theta)]^{\frac{1}{2}}} \right) \right\} .$$

We want to select positive weights p_ϑ with a sum that is bounded by 1 that permit the convergence to 0 of the right-hand side, as $z \to \infty$ and then $\tau \to \infty$.

In order to do so we split the bound into two parts, each of which involves a sum over a subset of parameters. One sum is over the region where $\|\vartheta - \theta\| >$

$(\log z/z)^{\frac{2}{\alpha}}$ and the other is over the region where $\tau z^{-\frac{2}{\alpha}} < \|\vartheta - \theta\| \leq (\log z/z)^{\frac{2}{\alpha}}$. For the first region we may take a uniform weight $p_\theta = 0.5 \cdot |T|^{-1} z^{-\frac{4}{\alpha}} \delta^2$.

A total uniform weight that is assigned to the first region is $1/2$. The cardinality of T is equal to $|T| z^{\frac{4}{\alpha}} \delta^{-2}$, where $|T| = (t_1 - t_0)(h_1 - h_0)$ is the area of T and δ is the increment used in the discrete approximation. Consequently, the magnitude of a weight is proportional to $z^{-\frac{4}{\alpha}}$, leading to $\log(p_\vartheta) \sim -(4/\alpha) \log z$. For all ϑ in the first region we have that

$$z^2 [1 - \mathrm{Cov}(Z_\vartheta, Z_\theta)] \gtrsim (\log z)^2 ,$$

which is more than enough to counteract the effect of the weight and to assure that the sum of the resulting normal probability is small.

Bounding the sum in the second region, which corresponds to a ring about θ, requires a more careful examination of the asymptotic expansion of the covariance. A typical parameter value in the second region may be represented as $\vartheta = \theta + (i\delta z^{-\frac{2}{\alpha}}, j\delta z^{-\frac{2}{\alpha}})$, for $\tau < |i|$ and $\tau < |j|$. One can find a positive constant c_0 such that

$$z^2 [1 - \mathrm{Cov}(Z_\vartheta, Z_\theta)] \geq c_0 z^2 \|\vartheta - \theta\|^\alpha = c_0 \delta^{\frac{\alpha}{2}} (|i|^2 + |j|^2)^{\frac{\alpha}{2}} ,$$

Replacing c_0 by a large enough finite constant c_1 produces an upper bound. If we choose

$$p_\vartheta = \epsilon^{-1} \exp \left\{ -(c_0/2)\delta^{\frac{\alpha}{2}} (|i|^2 + |j|^2)^{\frac{\alpha}{2}} \right\}$$

we get that, after letting $z \to \infty$,

$$\frac{\log(\epsilon p_\vartheta) + z^2 [1 - \mathrm{Cov}(Z_\vartheta, Z_\theta)][1 - \epsilon]}{\sqrt{2}z[1 - \mathrm{Cov}(Z_\vartheta, Z_\theta)]^{\frac{1}{2}}} \gtrsim \frac{c_0 \delta^{\frac{\alpha}{4}}}{\sqrt{8c_1}} \cdot (|i|^2 + |j|^2)^{\frac{\alpha}{4}}$$

$$= c \cdot (|i|^2 + |j|^2)^{\frac{\alpha}{4}} .$$

It follows that the sum of probabilities in the second region is asymptotically bounded by

$$\sum_{|i|>\tau} \sum_{|j|>\tau} \exp \left\{ - (c^2/2) \cdot (|i|^2 + |j|^2)^{\frac{\alpha}{2}} \right\} ,$$

which converges to zero when $\tau \to \infty$. This completes the validation of Condition I* in the Gaussian setting.

3.4.5 Checking Condition III*

The last condition to check is Condition III* that states the convergence of the expected ratio of the maximum over the sum and identifies the term that this limit approximates.

The ratio is a bounded functional of the local field. Therefore, in order to identify the limit of the expectation it is sufficient to identify the limit in distribution of the local field, as the threshold goes to infinity. The term that is approximated by the limit is obtained by increasing the range of the local field. In the current situation this corresponds to letting τ go to infinity.

In the case of sequential testing the local field involves partial sums. For $m > n$, where the distribution of X_i is determined by the null density f, the increments of the partial sum is of the form $\log\{g(X_i)/f(X_i)\}$. For $m < n$, on the other hand, the distribution of X_i is determined by the density g and the increments of the partial sum is of the form $-\log\{g(X_i)/f(X_i)\} = \log\{f(X_i)/g(X_i)\}$.

The distribution of the local field does not change with the change of the threshold. Therefore, the description fits the limit local field. The distribution is also independent of n, as long as $n_0 + \tau \le n \le n_1 - \tau$. Consequently, we may describe the term $E_n[\hat{M}_n/\hat{S}_n]$ by shifting the origin to be equal to n. Accordingly, the observations for $-\tau \le i < 0$ have density g and the observations for $0 < i \le \tau$ have density f. The partial sum for $m = 0$ is equal to 0. Partial sums are formed in one way for negative indices and in a different way for positive indices, covering the whole range $-\tau \le m \le \tau$.

The resulting partial sums are exponentiated and then either summed or maximized. The ratio between the maximum and the sum is the random variable that enters into the expectation. The term $E[\mathcal{M}/\mathcal{S}]$ is the outcome of the same process for $-\infty < m < \infty$. The fact that we have convergence as a function of τ results from the application of the relevant part of the proof that was used in order to establish Condition II*.

In the Gaussian setting the bounded random variables $\hat{M}_\theta/\hat{S}_\theta$ is produced by a finite collection of random variables with a multi normal distribution. This random variable converges in distribution, as $z \to \infty$, to the random variable that is produced as a function of the limit Gaussian distribution of the finite collection. This limit Gaussian distribution is determined by the limit expectation and the limit covariance structure. However, since

$$E_\theta[z(Z_\vartheta - Z_\theta)] = -z^2[1 - \text{Cov}(Z_\vartheta, Z_\theta)]$$

and

$$\text{Cov}[z(Z_\vartheta - Z_\theta), z(Z_\eta - Z_\theta)]$$
$$= z^2[1 + \text{Cov}(Z_\vartheta, Z_\eta) - \text{Cov}(Z_\vartheta, Z_\theta) - \text{Cov}(Z_\eta, Z_\theta)]$$

it is clear that these limits emerge from expanding the covariance of the original field about θ. This expansion is of one type when a smooth kernel is used and of a different type when the kernel is an indicator of an interval.

Consider the case of a smooth kernel g. We showed that in the case where $\text{Var}(Z_\theta) \equiv 1$ then the gradient of covariance, evaluated at $\vartheta = \theta$, is the zero vector. Consequently, the expansion is determined by higher order derivatives.

The fact that the variance is constant as a function of the parameters implies a relation between the Hessian and the gradient of the random field:

$$0 = \left(\frac{\partial}{\partial\vartheta}\right)^2 \text{Cov}(Z_\vartheta, Z_\vartheta) = 2\frac{\partial}{\partial\vartheta}\text{Cov}(\dot{Z}_\vartheta, Z_\vartheta)$$

$$= 2[\text{Cov}(\ddot{Z}_\vartheta, Z_\vartheta) + \text{Cov}(\dot{Z}_\vartheta, \dot{Z}_\vartheta)]$$

Therefore,

$$\left(\frac{\partial}{\partial\vartheta}\right)^2 \text{Cov}(Z_\vartheta, Z_\theta)|_{\vartheta=\theta} = \text{Cov}(\ddot{Z}_\theta, Z_\theta) = -\text{Cov}(\dot{Z}_\theta, \dot{Z}_\theta) = -\Sigma_\theta$$

where Σ_θ is the variance-covariance matrix of the gradient that was identified in Chapter 2: $\Sigma_\theta = \langle\dot{g}_\theta, \dot{g}_\theta\rangle/\|g\|^2$. In the special case where $g(x) = \exp\{-x^2/2\}$ this matrix became:

$$\Sigma_\theta = \begin{pmatrix} h^{-2} & 0 \\ 0 & \frac{1}{4}h^{-2} \end{pmatrix}.$$

Consider $\vartheta = \theta + (i\delta z^{-1}, j\delta z^{-1}) = \theta + \delta z^{-1}\iota$, where ι is a point on the two-dimensional grid of integers. The limit of the expectation is given by:

$$\lim_{z\to\infty} E_\theta[z(Z_\vartheta - Z_\theta)] = \lim_{z\to\infty} \frac{z^2}{2}(\vartheta - \theta)'\text{Cov}(\ddot{Z}_\theta, Z_\theta)(\vartheta - \theta) = -\frac{\delta^2}{2}\iota'\Sigma_\theta\iota$$

and the limit of the covariance between two elements associated with ϑ and η, where $\eta = \theta + \delta z^{-1}v$, is:

$$\lim_{z\to\infty} \text{Cov}[z(Z_\vartheta - Z_\theta), z(Z_\eta - Z_\theta)]$$

$$= \lim_{z\to\infty} \frac{z^2}{2}[(\vartheta - \theta)'\text{Cov}(\ddot{Z}_\theta, Z_\theta)(\vartheta - \theta) + (\eta - \theta)'\text{Cov}(\ddot{Z}_\theta, Z_\theta)(\eta - \theta)$$

$$- (\vartheta - \eta)'\text{Cov}(\ddot{Z}_\eta, Z_\eta)(\vartheta - \eta)]$$

$$= \frac{\delta^2}{2}[-\iota'\Sigma_\theta\iota - v'\Sigma_\theta v + (\iota - v)'\Sigma_\theta(\iota - v)] = \delta^2\iota'\Sigma_\theta v,$$

for ι and v two points on the grid of integers. The evaluation of the limit exploits the fact that $\Sigma_\eta \to \Sigma_\theta$. Recall that the expectation of $\dot{Z}_\theta$ is zero and the variance-covariance matrix is Σ_θ. We may conclude that the process $\{z(Z_\vartheta - Z_\theta)\}$ converges in distribution to the process:

$$\{\delta\dot{Z}'_\theta\iota - (\delta^2/2)\iota'\Sigma_\theta\iota : \iota = (i, j), |i| \leq \tau, |j| \leq \tau\},$$

defined over a square in the integer lattice.

The functional that measures the contribution of the local field, the expectation of the maximum of the exponentiated local field divided by its sum, converges to a functional that is defined in the same way but with respect to the limit local field:

$$E[\hat{\mathcal{M}}/\hat{\mathcal{S}}] = E\left[\frac{\max_{\|\iota\|\leq\tau}e^{\delta\dot{Z}'_\theta\iota-(\delta^2/2)\iota'\Sigma_\theta\iota}}{\sum_{\|\iota\|\leq\tau}e^{\delta\dot{Z}'_\theta\iota-(\delta^2/2)\iota'\Sigma_\theta\iota}}\right],$$

where the norm that is being used here is the $\|(x, y)\| = |x| \vee |y|$. However,

$$\delta\dot{Z}'_\theta\iota - (\delta^2/2)\iota'\Sigma_\theta\iota = -\frac{\delta^2}{2}(\iota - U)'\Sigma_\theta(\iota - U) + \frac{\delta^2}{2}\dot{Z}'_\theta\Sigma_\theta^{-1}\dot{Z}_\theta,$$

for $U = \Sigma_\theta^{-1}\dot{Z}_\theta$. Consequently,

$$E[\hat{\mathcal{M}}/\hat{\mathcal{S}}] = E\left[\frac{\max_{\|\iota\|\leq\tau}e^{-\frac{\delta^2}{2}(\iota-U)'\Sigma_\theta(\iota-U)}}{\sum_{\|\iota\|\leq\tau}e^{-\frac{\delta^2}{2}(\iota-U)'\Sigma_\theta(\iota-U)}}\right],$$

which converges to:

$$E[\mathcal{M}/\mathcal{S}] = E\left[\frac{\max_\iota e^{-\frac{\delta^2}{2}(\iota-U)'\Sigma_\theta(\iota-U)}}{\sum_\iota e^{-\frac{\delta^2}{2}(\iota-U)'\Sigma_\theta(\iota-U)}}\right],$$

with maximization and summation extending over the entire integer grid.

When expanding the covariance function in the case where the random field is not smooth one may see that the first-order terms do not vanish. In particular, in the scanning statistic example with an indicator serving as a kernel and the parameters are the end points of an interval, we obtained that:

$$1 - \text{Cov}(Z_\theta, Z_\vartheta) \approx \frac{1}{2h}|\vartheta_1 - \theta_1| + \frac{1}{2h}|\vartheta_2 - \theta_2| = \frac{\delta|i|}{2hz^2} + \frac{\delta|j|}{2hz^2},$$

for $\vartheta = \theta + (i\delta z^{-2}, j\delta z^{-2})$. It follows that:

$$\lim_{z\to\infty} E_\theta[z(Z_\vartheta - Z_\theta)] = -\frac{\delta|i|}{2h} - \frac{\delta|j|}{2h}.$$

If we consider the asymptotic covariance between the term associated with ϑ and the term associated with $\eta = \theta + (i'\delta z^{-2}, j'\delta z^{-2})$ we get:

$$\lim_{z\to\infty} \text{Cov}[z(Z_\vartheta - Z_\theta), z(Z_\eta - Z_\theta)]$$

$$= \frac{\delta}{2h}[|i| + |j| + |i'| + |j'| - |i - i'| - |j - j'|]$$

$$= \frac{\delta}{h}[|i| \wedge |i'| + |j| \wedge |j'|],$$

if i and i' (j and j') share the same sign. If i and i' (j and j') do not share the same sign then they make zero contribution to the covariance. One may associate the limit distribution of the random field $\{z(Z_\vartheta - Z_\theta)\}$ with a sum of two independent processes: $\{W_1(i\delta/h) + W_2(i\delta/h) : |i| \leq \tau, |j| \leq \tau\}$, where W_1 and W_2 are copies of the process $W(t) = B_t - |t|/2$. The process B_t is the double-ended Brownian motion which originates at 0 and evolves independently to the right and to the left of the origin.

The functional that measures the contribution of the local field converges in this case, due to the independence between the two processes, to:

$$E[\hat{\mathcal{M}}/\hat{\mathcal{S}}] = E\left[\frac{\max_{|i|,|j|\leq\tau} e^{W_1(i\delta/h)+W_2(j\delta/h)}}{\sum_{|i|\leq\tau}\sum_{|j|\leq\tau} e^{W_1(i\delta/h)+W_2(j\delta/h)}}\right] = \left\{E\left[\frac{\max_{|i|\leq\tau} e^{W(i\delta/h)}}{\sum_{|i|\leq\tau} e^{W(i\delta/h)}}\right]\right\}^2,$$

and when we send τ to infinity we get:

$$E[\mathcal{M}/\mathcal{S}] = \left\{E\left[\frac{\max_{|i|} e^{W(i\delta/h)}}{\sum_{|i|} e^{W(i\delta/h)}}\right]\right\}^2,$$

with maximization and summation extending to the entire collection of negative and positive integers.

3.5   Integration

The direction of analysis so far involved the exploration of finer and finer details until we were able to obtain an approximation of a relevant quantity associated with each given value of the parameter. Now we want to put things back together again in order to obtain the approximation of the tail probability of the extreme of the random field, which was our original motivation for the analysis.

The probability in question was presented as a sum over all parameter values of relevant quantities. The localization theorem produces an approximation for each quantity. In the final step we propose to replace the quantities in the sum by their approximated values and evaluate the resulting summation. It may be useful to think of the localization theorem as producing point-wise convergence for each value of the parameter. The integration step translates this point-wise convergence to convergence of integrals. Concepts of uniform integrability or theorems like the dominated convergence theorem may be useful in order to establish this translation.

Start with the problem of sequential testing. As a result of measure transformation we obtained the representation:

$$P\left(\max_{n\in T} \ell_n \geq x\right) = e^{-x} \sum_{n=n_0}^{n_1} E_n\left(\frac{M_n}{S_n} e^{-[\tilde{\ell}_n+m_n]}; \tilde{\ell}_n + m_n \geq 0\right).$$

The localization theorem informed us that

$$E_n\left(\frac{M_n}{S_n} e^{-[\tilde{\ell}_n+m_n]}; \tilde{\ell}_n + m_n \geq 0\right) \approx \frac{1}{\sqrt{n\sigma^2}}\phi\left(\frac{In-x}{\sqrt{n\sigma^2}}\right) E[\mathcal{M}/\mathcal{S}],$$

for some constant $E[\mathcal{M}/\mathcal{S}]$ that did not depend on n, for $n_0 + \tau \leq n \leq n_1 - \tau$. The same type of convergence holds for parameter values in the vicinity of the boundary points but with $E[\mathcal{M}/\mathcal{S}]$ replaced by a similar term that involves also the expected ratio of maximization and summation. However, the range of parameter values over which maximization and summation take place is partially truncated in either the positive or the negative direction.

If indeed the point-wise convergence implies convergence of the sum, we may conclude that:

$$P\left(\max_{n \in T} \ell_n \geq x\right) \approx e^{-x} E[\mathcal{M}/\mathcal{S}] \sum_{n=x/I-C\sqrt{x}}^{x/I+C\sqrt{x}} \frac{1}{\sqrt{n}\sigma} \phi\left(\frac{n - x/I}{\sqrt{n}\sigma/I}\right)$$

$$\approx e^{-x} \frac{E[\mathcal{M}/\mathcal{S}]}{I}[2\Phi(CI^{\frac{3}{2}}\sigma^{-1}) - 1],$$

with the second approximation resulting from the approximation of the sum by an integral, using in the process the fact that $n/x \approx I$.

The original problem dealt with the stopping time N_x and the probability that it obtains a finite value. This probability, the significance level of the test results from allowing $C \to \infty$, which produces:

$$\lim_{x \to \infty} e^x P(N_x < \infty) = \lim_{C \to \infty} \lim_{x \to \infty} e^x P\left(\max_{n \in T} \ell_n \geq x\right) = \frac{1}{I}E[\mathcal{M}/\mathcal{S}].$$

In Chapter 2 we obtained, applying different means, an approximation of the same probability. Comparing the terms of that approximation that is given in (2.3) with the current one, recalling that $I = E_g(\ell_1)$, teaches us that:

$$E[\mathcal{M}/\mathcal{S}] = \exp\left\{-\sum_{n=1}^{\infty} n^{-1}[P_g(\ell_n \leq 0) + P(\ell_n > 0)]\right\},$$

for the ratio between the maximized likelihood ratios and the sum of these likelihood ratios that emerges in the context of the current problem. In particular, if we deal with the problem of testing hypothesis for the normal mean being equal to 0 or to μ, we get the relation: $E[\mathcal{M}/\mathcal{S}] = (\mu^2/2) \cdot v(\mu)$. Interestingly enough, the log-likelihood ratios that are used for this testing problem can be represented in the form $\ell_n = \mu \sum_{i=1}^{n} X_i - n\mu^2/2 = B(\mu^2 n) - (\mu^2 n)/2 = W(\mu^2 n)$, with equality meaning equality in distribution. The conclusion we make is that

$$E\left[\frac{\max_{|i|} e^{W(i\mu^2)}}{\sum_{|i|} e^{W(i\mu^2)}}\right] = (\mu^2/2) \cdot v(\mu). \tag{3.2}$$

All this is nice, but we still need to show that point-wise convergence implies convergence of the sum. For that we may apply, for example, the Berry–Esseen theorem once more and obtain a universal constant such that

$$\mathrm{E}_n \left(\frac{M_n}{S_n} e^{-[\tilde{\ell}_n + m_n]}; \tilde{\ell}_n + m_n \geq 0 \right) \leq cn^{-\frac{1}{2}},$$

which gives us the convergence via the dominated convergence theorem.

Now is the time to turn back to the examination of the problem of scanning statistics. In the case of a smooth kernel we obtained:

$$\mathrm{P} \left(\max_{\theta \in T} Z_\theta \geq z \right) \approx \frac{e^{-\frac{1}{2} z^2}}{z \sqrt{2\pi}} \sum_{\theta \in T} \mathrm{E} \left[\frac{\max_\iota e^{-\frac{\delta^2}{2} (\iota - U)' \Sigma_\theta (\iota - U)}}{\sum_\iota e^{-\frac{\delta^2}{2} (\iota - U)' \Sigma_\theta (\iota - U)}} \right],$$

and in the case of an indicator of an interval serving as a kernel we got:

$$\mathrm{P} \left(\max_{\theta \in T} Z_\theta \geq z \right) \approx \frac{e^{-\frac{1}{2} z^2}}{z \sqrt{2\pi}} \sum_{\theta \in T} \left\{ \mathrm{E} \left[\frac{\max_{|i|} e^{W(i\delta/h)}}{\sum_{|i|} e^{W(i\delta/h)}} \right] \right\}^2.$$

Not only $\mathrm{E}[\mathcal{M}/\mathcal{S}]$ but also the parametrization and the discrete collection of parameters T is different for the two cases. For the smooth case we used a grid with span δz^{-1} and for the other case we used a grid with span δz^{-2}, defined for transformed parameters.

Approximating summation by integration we get for the smooth case that:

$$\lim_{z \to \infty} \frac{z}{z^2} e^{\frac{1}{2} z^2} \mathrm{P} \left(\max_{\theta \in T} Z_\theta \geq z \right) \approx \frac{1}{\delta^2 \sqrt{2\pi}} \int_T \mathrm{E} \left[\frac{\max_\iota e^{-\frac{\delta^2}{2} (\iota - U)' \Sigma_\theta (\iota - U)}}{\sum_\iota e^{-\frac{\delta^2}{2} (\iota - U)' \Sigma_\theta (\iota - U)}} \right] d\theta,$$

where now T corresponds to the original continuous collection of parameters $T = [t_0, t_1] \times [h_0, h_1]$. If we allow $\delta \to 0$ we will get that the maximum on the right-hand side converges to 1 and the sum, upon multiplication by δ^2, converges to a Gaussian integral. The result in the smooth case is the approximation:

$$\lim_{z \to \infty} z^{-1} e^{\frac{1}{2} z^2} \mathrm{P} \left(\max_{\theta \in T} Z_\theta \geq z \right) = (2\pi)^{-3/2} \int_T |\Sigma_\theta|^{\frac{1}{2}} d\theta,$$

that is in agreement with (2.5).

In the non-smooth case we exploit the fact that the transformation of the parameters is measure-preserving and the representation of the expectation of the maximum over the sum in terms of the overshoot function ν of the normal random walk. The approximation of the sum by an integral becomes:

$$\lim_{z \to \infty} \frac{z}{z^4} e^{\frac{1}{2} z^2} \mathrm{P} \left(\max_{\theta \in T} Z_\theta \geq z \right) = \frac{1}{\delta^2 \sqrt{2\pi}} \int_T \left[(\delta/2h) \cdot \nu \left((\delta/h)^{\frac{1}{2}} \right) \right]^2 d\theta.$$

When $\delta \to 0$ the ν function converges to 1. Therefore,

$$\lim_{z \to \infty} z^{-3} e^{\frac{1}{2} z^2} \mathrm{P} \left(\max_{\theta \in T} Z_\theta \geq z \right) = (2\pi)^{-\frac{1}{2}} \cdot (0.5)^2 (t_1 - t_0)(1/h_0 - 1/h_1),$$

which is identical to the approximation in (2.10) that resulted from the application of the double-sum method.

The last point to check is uniform integrability. This follows from using the bound $z^{-1}(2\pi)^{-\frac{1}{2}}$ on the density of $z(Z_\theta - z)$ in both versions of Gaussian scanning statistics. This completes the proof of the convergence of the sum in the scanning statistic example.

In summary, in this chapter using our unifying method we were able to prove again the results from the previous chapter regarding approximations of tail distributions of some random fields. In all the examples the asymptotic derivation involved probabilities that converge to zero. In the next chapter we will generalize the results and deal with probabilities that do not converge to zero.

4

From the local to the global

4.1 Introduction

In Chapter 3 we used a method for the approximation of the tail probability of extremes in a random field. The method was applied in order to analyze two examples: an example associated with sequential testing and an example associated with scanning statistics. This method is meaningful in settings of large deviation where the probability in question converges to zero. The approximation produced by the method involved the summation, over the set of parameters, of terms. The terms were a product of a large deviation factor and two other factors that correspond to lower-order contributions. Technically, the statement of large deviation for the probability may be translated to a relation between the the large deviation factors, one for each term in the sum, which are exponentially or super-exponentially small as a function of the threshold, and the integration of such factors, which accumulates the small terms together. The setting that is assumed in the approximation is such that, even after the accumulation of all terms, the sum still converges to zero.

In the next section we deal with the case where the parameter space is large enough, in comparison with the threshold, to make the probability of crossing the threshold converge to a positive number. We approach this problem by the introduction of the Poisson approximation as a vehicle to integrate small probabilities of weakly dependent events. This Poisson approximation resembles in spirit the double-sum argument in the double-sum method.

Occasionally, one may be interested in settings where the probability of crossing the threshold is 1 and in functionals other than the probability. For example, in sequential change-point detection, where a change-point detection monitoring statistic is followed over time, one may be interested in the expected duration of time that elapsed until the first crossing of the detection threshold. Subject to a Poisson approximation, an approximation for the expectation will emerge

Extremes in Random Fields: A Theory and its Applications, First Edition. Benjamin Yakir.
© 2013 by Higher Education Press. All rights reserved. Published 2013 by John Wiley & Sons, Ltd.

as a corollary of proving uniform integrability. We demonstrate this claim in Section 4.3 where we present the cumulative sum (cusum) method for change-point detection, an adaptation of the sequential probability ratio test to the setting of change-point detection.

4.2 Poisson approximation of probabilities

The example of a scanning statistic involved the examination of the region $T = [t_0, t_1] \times [h_0, h_1]$, with $[t_0, t_1]$ an interval of the real line where the signal may be located and $[h_0, h_1]$ a range of parameter values that characterizes the width of a signal. A kernel function was used in order to construct the monitoring statistic. In the case where we employed the kernel $g(x) = \exp\{-x^2/2\}$ we obtained the approximation:

$$P\left(\sup_{\theta \in T} Z_\theta \geq z\right) \approx z e^{-\frac{1}{2}z^2} (2\pi)^{-3/2} \cdot 0.5 \cdot (t_1 - t_0)(1/h_0 - 1/h_1)$$

for the probability of crossing the threshold z within the region T. If the kernel $g(x)$ is the indicator of the interval $[-0.5, 0.5]$ then the resulting approximation was:

$$P\left(\max_{\theta \in T} Z_\theta \geq z\right) \approx z^3 e^{-\frac{1}{2}z^2} (2\pi)^{-1/2} \cdot (0.5)^2 \cdot (t_1 - t_0)(1/h_0 - 1/h_1) .$$

Both approximations were constructed having in mind the situation where the parameter set is kept fixed in size and the threshold z is allowed to go to infinity.

If, as may be the case, the interval $[t_0, t_1]$ is very large one may question the appropriateness of an asymptotic expansion that considers the length of the interval to be a fixed quantity. More meaningful results may be obtained by imagining that the length of the interval is diverging to infinity together with the threshold. Examination of the proof that we used reveals that the approximation is still appropriate when the length of the interval is diverging to infinity at a modest rate that does not prevent an overall convergence to zero of the approximation. However, if the divergence is fast enough, leading to a non-vanishing probability of crossing the threshold, then the arguments that were used in order to establish the approximation may no longer be valid. Instead, a new method is needed in order to deal with this new problem.

In a way, the issue is not a new one and we have already encountered a similar problem when we discussed the double-sum method. Specifically, in the double-sum method, in order to analyze the probability of crossing the threshold within a fixed region T by a statistic that employs the indicator of an interval as the kernel, we subdivided the region into smaller sub-regions of the form:

$$T_{ij} = T_\theta = \{\vartheta = (\vartheta_1, \vartheta_2) : 0 \leq \vartheta_1 - \theta_1 \leq \tau/z^2, 0 \leq \vartheta_2 - \theta_2 \leq \tau/z^2\},$$

for $\theta = \theta_{ij}$ in a two-dimensional grid with span τ/z^2. We were able to produce an approximation for the probability of crossing in each such sub-region. This approximation was extended to the entire region via the double-sum argument that established the fact that the sum, over the grid, of the probabilities of crossing within a sub-interval is an approximation to the probability of the union, i.e., the probability of crossing in T.

In the current problem we also have a method for dealing with relatively small sub-regions and we want to extend the outcomes that are obtained for smaller sub-regions to a larger region. Using the same line of argument, one may propose to subdivide the increasing region $T = [t_0, t_1] \times [h_0, h_1]$ into disjoint sub-regions of fixed or modestly increasing volume. The number of such sub-regions is increasing fast with the threshold but not the volume of each sub-region. Since we have a valid approximation for the probability associated with each sub-region we may be able to use the double-sum argument in order to produce an approximation that is based on the sum of approximations of the sub-regions probabilities. Indeed, this approach may be put to work in the current problem. However, in order to make it work we need a slightly more sophisticated methodology, the methodology of the Poisson approximation.

The Poisson random variable is associated with the counting of the number of occurrences of rare events. The distribution of the count is characterized by λ, the expected number of occurrences. Hence, if W is the total number of occurrences and is Poisson then

$$P(W = k) = e^{-\lambda}\frac{\lambda^k}{k!}, \quad k = 0, 1, 2, \ldots .$$

In particular, $P(W = 0) = e^{-\lambda}$ is the probability of no occurrence and $P(W > 0) = 1 - e^{-\lambda}$ is the probability of at least one occurrence.

In order to associate the Poisson approximation with the problem at hand we need to specify counts. Consider sub-regions of the form $T_i = [t_0 + (i - 1)m, t_0 + im] \times [h_0, h_1]$, for $i = 1, 2, \ldots (t_1 - t_2)/m$. An example of such a split for the scanning statistic that uses an indicator as a kernel is given in Figure 4.1. Each point in the parameter space is associated with an interval of observations. Several examples of parameter points and their respective intervals are given. Notice that whenever two points are separated by at least one sub-region then the intervals that are associated with them cannot overlap. The most extreme examples that are associated with the longest and closest intervals are plotted for T_1 and T_3. Still, the associated intervals do not intersect.

Let X_i be the indicator of an event that at least one crossing of the threshold occurred within the sub-region T_i:

$$\{X_i = 1\} = \left\{ \sup_{\theta \in T_i} Z_\theta \geq z \right\}. \tag{4.1}$$

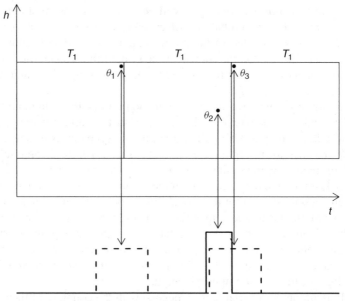

*Figure 4.1   A partition of the parameter set into disjoint sub-regions $T_1, T_2, \ldots$.
Parameters are associated with intervals of observations when the indicator
kernel is used. Intervals that are associated with neighboring regions may overlap. However, intervals that are associated with non-neighboring regions cannot
overlap.*

The sum of indicators given by $\hat{W} = \sum_{i=1}^{(t_1-t_2)/m} X_i$ is a count. This sum is equal
to 0 if, and only if, all indicators in the sum are equal to zero. It follows that:

$$\left\{ \sup_{\theta \in T} Z_\theta \geq z \right\} = \left\{ \max_{1 \leq i \leq (t_1-t_0)/m} X_i = 1 \right\} = \{ \hat{W} > 0 \} .$$

Thus, we can assess the probability of crossing the threshold by considering the
asymptotic distribution of $\hat{W}$. We proceed by approximating this distribution.

The problem that involves an interval kernel is slightly more simple to analyze
so we approach it first. Notice that the probability of the event indicated by X_i
is proportional to $mz^3 e^{-\frac{1}{2}z^2}$, thus converging to 0 when m is sub-exponential.
There are $(t_1 - t_2)/m$ such events. If the length of the scanning region, $t_2 - t_1$,
is asymptotically proportional to $z^{-3} e^{\frac{1}{2}z^2}$ then the sum of probabilities converges
to a constant:

$$\lambda = \lim_{z \to \infty} (t_1 - t_0) z^3 e^{-\frac{1}{2}z^2} (2\pi)^{-1/2} \cdot (0.5)^2 \cdot (1/h_0 - 1/h_1) ,$$

for $0 < \lambda < \infty$.

The limit λ is the asymptotic expectation of $\hat{W}$. The standard Poisson approximation of the binomial distribution states that if the events that are indicated by

the X_i's were independent then the distribution of $\hat{W}$ would have converged to the Poisson distribution with expectation λ. The proof of the statement is simple and relies on a simple analytic expansion of binomial probabilities. Unfortunately, in the current example the events are not quite independent so the standard argument for proving the Poisson approximation is not sufficient. However, extensions of the theorem to settings that involve local dependencies between events do exist and enable us nonetheless to obtain the same Poisson limit that was predicted under the (false) assumption of independence.

A very useful formulation of a general Poisson limit theorem is given in [9]. This theorem considers a countable collection of zero-one indicator variables X_i indexed by the elements of I. The total variation distance between the sum of indicator variables and the Poisson distribution, with the same expectation as the sum, is bounded in the theorem by a sum of three terms. These three terms are constructed in relation to 'neighborhoods of dependence' I_i, associated with each element in the sum. The first of the three terms measures the total size of these neighborhoods of dependence and is denoted b_1. The second term measures the strength of dependence between the elements within neighborhoods and is denoted b_2. The third term measures the total dependence between each element and elements outside its neighborhood of dependence. It is denoted b_3. Specifically, the terms are:

$$b_1 = \sum_{i \in I} \sum_{j \in I_i \setminus \{i\}} P(X_i = 1)P(X_j = 1) \,,$$

$$b_2 = \sum_{i \in I} \sum_{j \in I_i \setminus \{i\}} P(X_i = 1, X_j = 1) \,,$$

$$b_3 = \sum_{i \in I} E\left[|E(X_i | \sigma\{X_j : j \notin I_i\}) - E(X_i)|\right] \,,$$

where $\sigma\{X_j : j \notin I_i\}$ is the σ-algebra generated by the collection $\{X_j : j \notin I_i\}$.

From the statement of the theorem one may conclude that:

$$|P(\hat{W} > 0) - (1 - e^{-\lambda})| \leq b_1 + b_2 + b_3 + |e^{-E(\hat{W})} - e^{-\lambda}|.$$

Consequently, a demonstration that the three terms b_1, b_2, and b_3 can be constructed in a way that make them converge to 0, as $z \to \infty$, will produce the approximation:

$$P\left(\max_{\theta \in T} Z_\theta \geq z\right) \approx 1 - \exp\left\{-z^3 e^{-\frac{1}{2}z^2} (2\pi)^{-1/2} \cdot (0.5)^2 \cdot (t_1 - t_0)(1/h_0 - 1/h_1)\right\}.$$

$$(4.2)$$

This approximation covers the situations where the probability does not converge to zero. If the probability does converge to zero then this approximation and the original one are asymptotically equivalent.

In order to produce a representation that can be used in the theorem let $m > h_1/2$ and define the neighborhood of dependence of the indicator X_i to be $I_i = \{i - 1, i, i + 1\}$ [with obvious modifications for $i = 1$ and $i = (t_1 - t_0)/m$]. Observe that X_i is a function of the Gaussian white noise in the range $[t_0 + (i - 1)m - h_1/2, t_0 + im + h_1/2]$. It follows that X_i is independent of $\{X_j : j \notin I_i\}$ and, as a result, $b_3 = 0$.

All the probabilities that are involved in the computation of b_1 are equal to each other. Therefore,

$$b_1 = [2(t_0 - t_1)/m - 4]\{P(X_1 = 1)\}^2 + 2P(X_1 = 1) ,$$

the last summand being associated with the two edge elements. It follows that b_1 is asymptotic to $[2\lambda + 2]P(X_1 = 1)$, which converges to zero since $P(X_1 = 1)$ does.

Likewise, for b_2 we have:

$$b_2 = [2(t_0 - t_1)/m - 2]P(X_1 = 1, X_2 = 1) .$$

Redefine parameter sub-regions:

$$T_1 = [t_0, t_0 + m - h_1/2] \times [h_0, h_1] ,$$
$$T_2 = [t_0 + m - h_1/2, t_0 + m + h_1/2] \times [h_0, h_1] ,$$
$$T_3 = [t_0 + m + h_1/2, t_0 + 2m] \times [h_0, h_1] ,$$

and let the variables Y_i, , $i = 1, 2, 3$, be defined via $\{Y_j = 1\} = \{\max_{\theta \in T_i} Z_\theta \geq z\}$, as the indicators of crossing the threshold in the appropriate sub-regions. Notice that the indicators Y_1 and Y_3 are independent of each other and they share the same distribution. We use the fact that unless crossing occurs in a shared sub-region it must simultaneously occur in two disjoint sub-regions in order to have double crossing. As a consequence of this fact we obtain the upper bound $X_1 \cdot X_2 \leq Y_2 + Y_1 \cdot Y_3$ and can claim that:

$$P(X_2 = 1, X_3 = 1) \leq P(Y_2 = 1) + \{P(Y_1 = 1)\}^2 \leq P(Y_2 = 1) + \{P(X_1 = 1)\}^2 .$$

The probability $P(Y_2 = 1)$ is proportional to $h_1 z^3 e^{-\frac{1}{2}z^2}$. Consequently, b_2 is asymptotically bounded by $2\lambda[h_1/m + P(X_1 = 1)]$. Hence, b_2 converges to zero whenever $m \to \infty$. This completes the proof of the validity of approximation (4.2).

For the case of the smooth kernel $g(x) = \exp\{-x^2/2\}$ we would like to establish the generalizing approximation:

$$P\left(\max_{\theta \in T} Z_\theta \geq z\right) \approx 1 - \exp\{-z e^{-\frac{1}{2}z^2}(2\pi)^{-3/2} \cdot 0.5 \cdot (t_1 - t_0)(1/h_0 - 1/h_1)\} .$$

$$(4.3)$$

The difficulty is that in this case any pair of elements of the field, Z_θ and Z_ϑ, are correlated regardless of how far ϑ is from θ. Consequently, b_3, the most difficult to analyze of the three terms that bound the error in the Poisson approximation, is not equal to zero and cannot be ignored.

Instead of analyzing the term b_3 we will use a truncation argument in order to construct a field $\{\hat{Z}_\theta : \theta \in T\}$ that approximates the original field and yet has the property that remote elements are independent of each other.

Set the smooth kernel

$$\hat{g}(x) = \begin{cases} \exp\{-x^2/2\} & |x| \leq y, \\ \exp\{-x^2/2\}f(|x|-y) & |x| > y, \end{cases}$$

where f is a smooth and monotone decreasing function that is equal to 1 at zero and equal to 0 for all values larger than 1. Let $\hat{Z}_\theta$ be the standardized statistics that is constructed with $\hat{g}$ as the kernel. Since the kernel is smooth the formulae (2.5) for smooth Gaussian field apply. Note that for the approximating field any two elements with locations that are more than $2h_1(y+1)$ apart are independent of each other.

The approximating field $\hat{Z}_t$ is related to the original field via:

$$Z_\theta - \frac{\|\hat{g}\|}{\|g\|}\hat{Z}_\theta = \frac{1}{h^{\frac{1}{2}}\|g\|}\int_{\{|x-t|>hy\}}[1 - f(|x-t|/h - y)]e^{-\frac{(x-t)^2}{2h^2}}\,dB_x = \Delta_\theta .$$

Consequently, we may bound the probability that we seek to approximate both from above and from below using the approximating field and the associated field of discrepancies:

$$P\left(\max_{\theta \in T} Z_\theta \geq z\right) \leq P\left(\max_{\theta \in T} \hat{Z}_\theta \geq \frac{\|g\|}{\|\hat{g}\|}\left[z - \frac{\epsilon}{z}\right]\right) + P\left(\max_{\theta \in T} |\Delta_\theta| \geq \frac{\epsilon}{z}\right),$$

$$P\left(\max_{\theta \in T} Z_\theta \geq z\right) \geq P\left(\max_{\theta \in T} \hat{Z}_\theta \geq \frac{\|g\|}{\|\hat{g}\|}\left[z + \frac{\epsilon}{z}\right]\right) - P\left(\max_{\theta \in T} |\Delta_\theta| \geq \frac{\epsilon}{z}\right).$$

We proceed by showing that the probabilities associated with $\hat{Z}_\theta$ may be approximated using (4.3) and the probabilities associated with the discrepancies Δ_θ are vanishingly small.

For the probabilities that are associated with $\hat{Z}_\theta$ we may use the Poisson approximation. In particular, if m and y are selected so that $y/m \to 0$ then the term b_3 may be eliminated. The approximations will (almost) coincide with (4.3) if $\|g\|/\|\hat{g}\| = 1 + o(z^{-1})$ and if the matrix $|\hat{\Sigma}_\theta|$, the determinant of the matrix of inner products of the components of the gradient of $\hat{g}$, converges to $|\Sigma_\theta| = h^{-4}/4$. However,

$$0 \leq \|g\|^2 - \|\hat{g}\|^2 = 2\int_y^\infty [1 - f(x-y)]^2 e^{-x^2}\,dx \leq \frac{\sqrt{2}}{y}e^{-y^2} .$$

Therefore, provided that $(\log z)^{\frac{1}{2}}/y \to 0$ we have the required rate of convergence. A similar argument, applied to the gradient with respect to θ of $\hat{g}_\theta$, will show the convergence of the determinant whenever $y \to \infty$.

Let

$$\sigma^2 = \max_{\theta \in T} \mathrm{Var}(\Delta_\theta) = \frac{2}{\|g\|^2} \int_y^\infty [1 - f(x - y)]^2 e^{-x^2} dx \leq \frac{2}{\sqrt{\pi} y} e^{-y^2}.$$

We use Fernique's inequality (Theorem 3.1) with $\varphi(x) = c_1 x^2$, for an appropriate c_1, and choose $\lambda = \sqrt{\sigma}$. The inequality produces the bound

$$\mathrm{P}\left(\max_{\theta \in T} |\Delta_\theta| \geq \frac{\epsilon}{z}\right) \leq B\sigma^{-1}(t_2 - t_1)(h_1 - h_0)e^{-C\epsilon^2/(z^2\sigma^2)},$$

for some finite constants B and C. The order of magnitude of the right-hand side is $(y/z)\exp\{y^2 + z^2/2 - \sqrt{\pi}C\epsilon^2 z^{-2} y e^{y^2}/2\}$, which converges to zero for $y = \log z$. Thus, selecting y to be this large will be sufficient in order to validate (4.3).

4.3   Average run length to false alarm

We would like to discuss the Poisson approximation in the sequential setting. However, the probability that we computed in the context of the sequential probability ratio test converged to 0 for a maximization that already involved the entire parameter space. Consequently, a Poisson approximation will not add new information. Instead, we will consider a closely related problem of sequential monitoring that involves statistics that resemble the statistics that are used for the sequential probability ratio test.

The problem that we would like to consider emerged originally in the context of industrial production quality control. In the third decade of the last century Shewhart invented the control charts that are named after him. In these charts a quality index is plotted sequentially. As long as the index is below a pre-specified threshold the production is judged to be in control. If the threshold is crossed then the presumption is that something went wrong in the production line prompting action to restore quality. Thirty years later, Page suggested to use the cusum statistic as the quality index. Our goal will be to investigate the statistical properties of Page's cusum statistic, expressed in the same formulation that was used for the sequential probability ratio test.

Consider the construction of the quality index based on the accumulation of n independent observations $X_1, X_2, \ldots, X_n$. The distribution of observation X_i is f if production was in control at the ith production period and it is g if it went out of control. If production was in control for the entire duration then the density of all observations is f. On the other hand, if production was in control prior to the kth production period, but then something went wrong causing the kth and subsequent observations to have the alternative density g then we will get a likelihood which involved $k - 1$ products of f followed by $n - k + 1$ products of g. The log-likelihood ratio statistic to compare the null in-control distribution to an alternative that involves a change at time k is given by:

$$\ell_{k,n} = \sum_{i=k}^{n} \log \{g(X_i)/f(X_i)\} .$$

The change can occur at any production period and is treated as an unknown parameter. Estimating this unknown parameter from the data, using the maximum-likelihood estimator, produces the cusum statistic:

$$q_n = \max_{1 \leq k \leq n} \ell_{k,n}$$

that serves as the quality index. The stopping time that is associated with this index corresponds to the first time that the quality index crosses the threshold:

$$N_x = \inf \{n : q_n \geq x\} .$$

We are interested in the distribution of the stopping rule N_x when the process is in control.

We start with the investigation of the probability $P(N_x \leq m)$, which is carried out by a reformulation of the event in terms of a random field:

$$\{N_x \leq m\} = \left\{ \max_{1 \leq n \leq m} q_n \geq x \right\} = \left\{ \max_{1 \leq n \leq m} \max_{1 \leq k \leq n} \ell_{k,n} \geq x \right\} .$$

This formulation is similar to the formulation that was used for the sequential probability ratio test, but there are differences. The main difference is that in the current example we have a two-dimensional random field, parameterized by the pair (k, n) in the triangle $1 \leq k \leq n \leq m$, and not a one-dimensional process (see Figure 4.2). This will introduce some modifications in the analysis and will produce a different asymptotic behavior. However, most of the details that are involved in the application of the advocated method are more or less the same. Consequently, we will allow ourselves the freedom to skip the proof of some of the details. (This freedom is not granted to the active reader. Please check to see that you agree with all the statements that are made and you know how to complete the proofs, especially the conditions of Theorem 5.2.)

The first step is localization. Observe that the marginal probabilities of crossing the threshold x are a function of the difference $n - k + 1$ and are maximized when $n - k + 1 = x/I$. We can restrict the maximization to the strip:

$$T = \left\{(k, n) : x/I - C\sqrt{x} \leq n - k + 1 \leq x/I + C\sqrt{x}\right\} ,$$

for a large C, without compromising the probability by much. This region is indicated in Figure 4.2 using vertical lines. You may use the fact that the probability that we seek to approximate is of the order me^{-x}. Consequently, the argument that was used for localization in the case of the probability ratio test can be

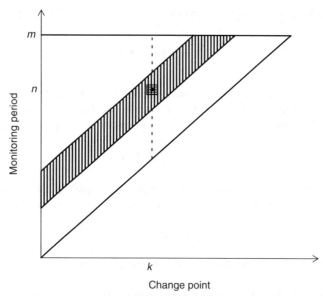

Figure 4.2   The parameter space for the cusum statistic. The original set of parameters is the upper triangle that is restricted by m. The first localization step reduces the parameters to the diagonal strip shaded with vertical lines. This reduction is obtained by showing for a typical k parameter that the parameter values that are indicated by a broken line make only a negligible contribution. After the application of the likelihood ratio identity the set is further reduced by the localization theorem. For a typical pair (k, n) indicated by a point the local region is the square about that point. The square is shaded with horizontal lines.

carried out for each k, $1 \leq k \leq m$. (For clarity, the values of parameters that are associated with a given k and are removed in the first localization step are indicated in Figure 4.2 by a broken line.) The resulting errors can be summed up in order to produce a bound on the overall error. The bound on the error will be of a smaller order compared with the size of the probability.

The second step is the transformation of the measure via the likelihood ratio identity. Here we use the likelihood ratio $\sum_{(k,n)\in T} \exp\{\ell_{k,n}\}$ to get the representation:

$$P\left(\max_{(k,n)\in T} \ell_{k,n} \geq x\right) = e^{-x} \sum_{(k,n)\in T} \mathrm{E}_{k,n}\left(\frac{M_{k,n}}{S_{k,n}} e^{-[\tilde{\ell}_{k,n}+m_{k,n}]}; \tilde{\ell}_{k,n} + m_{k,n} \geq 0\right),$$

where $S_{k,n} = \sum_{(j,h)\in T} e^{\ell_{j,h}-\ell_{k,n}}$ and $M_{k,n} = \max_{(j,h)\in T} e^{\ell_{j,h}-\ell_{k,n}}$ are the sum, respectively the maximal, of likelihood ratios, $m_{k,n} = \log M_{k,n}$, and $\tilde{\ell}_{k,n} = \ell_{k,n} - x$.

For the application of the localization theorem we use $\kappa = n - k + 1$ and take $\mathcal{F}_\kappa = \sigma\{X_{k-\tau}, \ldots, X_{k+\tau}, X_{n-\tau}, \ldots, X_{n+\tau}\}$, for the observations in the vicinity of either k or n. In particular, this σ-field is independent of the observations $X_{k+\tau+1}, \ldots, X_{n-\tau-1}$. This local region is marked in Figure 4.2 by horizontal lines and it produces a small square about the point (k, n).

The local field $\{\ell_{j,h} - \ell_{k,n}\}$, restricted to the region $|h - n| \leq \tau$ and $|j - k| \leq \tau$, is measurable with respect to the given σ-algebra. We produce $\hat{M}_\kappa$ and $\hat{S}_\kappa$ by restricting the maximization (respectively, summation) to the reduced region of parameter values. The distribution of the ratio $\hat{M}_\kappa/\hat{S}_\kappa$ is independent of κ, for all pairs (k, n) not too close to the boundary of the strip T. Moreover, this ratio can be written as a product of two independent ratios of maxima of an exponentiated one-dimensional local field divided by the sum of of the exponentiated local field. One ratio is associated with the observations in the vicinity of k and the other is associated with the observations in the vicinity of n. The ratios are independent of each other and each of them is of the same structure as the ratio obtained in the analysis of the sequential probability ratio test. It follows that

$$\lim_{\tau \to \infty} \mathrm{E}_{k,n}[\hat{M}_{k,n}/\hat{S}_{k,n}] = \{\mathrm{E}(\mathcal{M}/\mathcal{S})\}^2 \,,$$

where $\mathrm{E}(\mathcal{M}/\mathcal{S})$ is the limit obtained for the probability ratio test.

The conclusion from the application of the localization theorem is that

$$\mathrm{P}\left(\max_{(k,n)\in T} \ell_{k,n} \geq x)\right)$$

$$\approx e^{-x} \sum_{(k,n)\in T} \frac{1}{\sqrt{(n-k+1)\sigma^2}} \phi\left(\frac{I(n-k+1)-x}{\sqrt{(n-k+1)\sigma^2}}\right) \{\mathrm{E}[\mathcal{M}/\mathcal{S}]\}^2 \,,$$

where σ^2 is the variance of a log-likelihood ratio of an observation under the alternative distribution and ϕ is the standard normal density. Finally, the integration step, the approximation becomes:

$$\lim_{x \to \infty} (e^x/m)\mathrm{P}\left(\max_{(k,n)\in T} \ell_{k,n} \geq x\right) = \frac{1}{I}\{\mathrm{E}[\mathcal{M}/\mathcal{S}]\}^2[2\Phi(CI^{\frac{3}{2}}\sigma^{-1}) - 1] \,.$$

The extra factor of m results from the fact that each value of $n - k + 1$ appears approximately m times in the strip T. Letting $C \to \infty$ results in the statement:

$$\lim_{x \to \infty} (e^x/m)\mathrm{P}(N_x \leq m) = \frac{1}{I}\{\mathrm{E}[\mathcal{M}/\mathcal{S}]\}^2 \,. \tag{4.4}$$

In the special case of monitoring for a change in a normal mean from 0 to μ the constant on the right-hand side of (4.4) becomes $(\mu^2/2) \cdot \{\nu(\mu)\}^2$, for ν the function associated with an overshoot in a stopped normal random walk.

The approximation in (4.4) is valid as long as $x \ll m \ll e^x$, which makes the probability converge to zero. Approximation of the entire distribution of N_x calls for the use of a Poisson approximation.

An attempt to apply the Poisson approximation directly on N_x will result in a nonzero b_3 term in the bound on the approximation. In order to avoid this we use a truncation argument. The original stopping time is defined with the aid of the quality index $q_n = \max_{1 \le k \le n} \ell_{k,n}$. We consider instead an approximation of the quality index that maximizes the log-likelihood ratios only in the range that matters. Fixing a large C, let

$$\hat{q}_n = \max_{\{k:|n-k-I/x| \le C\sqrt{x}\}} \ell_{k,n}$$

and define $\hat{N}_x = \inf\{n : \hat{q}_n \ge x\}$. We will start by analyzing the distribution of $\hat{N}_x$ and then show that for a large value of C the distribution of $\hat{N}_x$ and the distribution of N_x are about the same.

Consider the time interval $[1, ye^x]$ and define $T = \{(k, n) : 0 < n \le ye^x, |n - k - I/x| \le C\sqrt{x}\}$. Clearly,

$$P(\hat{N}_x \le ye^x) = P\left(\max_{(k,n) \in T} \ell_{k,n} \ge x\right).$$

In order to analyze the probability on the right-hand side we divide T into disjoint regions:

$$T_i = \{(k, n) : (i-1)m < n \le im, |n - k - I/x| \le C\sqrt{x}\}, \quad i = 1, 2 \ldots, ye^x/m.$$

A partition of T to disjoint regions is demonstrated in Figure 4.3. Observe that restricted subset of parameters that is associated with $\hat{N}_x$ is associated with a diagonal strip. The strip is shaded with vertical lines. This strip is partitioned with bold, horizontal lines into subsets. These are regions T_i.

Define indicators X_i via the relation $\{X_i = 1\} = \{\max_{(k,n) \in T_i} \ell_{k,n} \ge x\}$. The sum of these indicators, $\hat{W} = \sum_{i=1}^{ye^x/m} X_i$ is the random variable that we want to approximate by the Poisson distribution. The expectation of the limit distribution is the limit of the sum of probabilities of the indicators:

$$\lambda = \lim_{x \to \infty} y(e^x/m)P(X_1 = 1) = \frac{y}{I}\{E[\mathcal{M}/\mathcal{S}]\}^2[2\Phi(CI^{\frac{3}{2}}\sigma^{-1}) - 1].$$

As a justification for the limit we may argue that apart from minor edge effects for X_1 the distribution of all the indicators is the same. In any case, the rate at which their probability converges to zero when $x \to \infty$, and for $x \ll m \ll e^x$ is the same and is given by the rate that was computed for X_1.

In order to validate the Poisson convergence of $\hat{W}$ we need to define neighborhoods of dependence and show that the terms b_1 and b_2 converge to zero. As before we set $I_i = \{i - 1, i, i + 1\}$, $1 < i < ye^x/m$ and let the neighborhoods on the edges contain two indexes only. By construction, X_i is independent of $\{X_j : j \notin I_i\}$ so indeed $b_3 = 0$.

To clarify that point consult Figure 4.3 again. An extreme parameter point is marked and the two endpoints of the periods of inspection that are associated with this parameter point are marked on the y-axis. This interval may share inspection

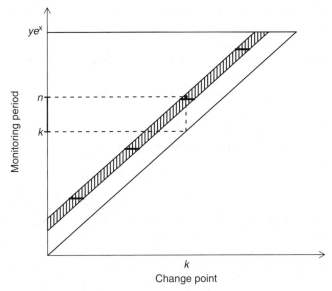

Figure 4.3   A partition of the parameter set into disjoint sub-regions. The range of monitoring periods that are associated with the parameter indicated by a point is marked as an a interval in bold on the y-axis.

periods with parameter values that belong to the sub-region immediately before the sub-region that contains (k, n). However, there are no monitoring periods common with parameter points that belong to the earlier sub-region. A parallel situation can be observed at the other end of the given sub-region. Observations may be shared with the subsequent sub-region but not with those that follow. Each time that parameters are separated by at least one sub-region then their respective log-likelihoods share no observations.

An argument no different than the argument that was used for the scanning statistic will produce:

$$b_1 = [2ye^x/m - 4]\{P(X_1 = 1)\}^2 + 2P(X_1 = 1) ,$$

which is asymptotic to $[2\lambda + 2]P(X_1 = 1)$ and converges to 0. For b_2 we have:

$$b_2 = [2ye^x/m - 2]P(X_1 = 1, X_2 = 1) .$$

Let w be such that $x \ll w$ but $w/m = o(1)$.

Splitting the union $T_1 \cup T_2$ into three disjoint subsets, according to $n < m - w$, $m - w \le n \le m + w$ and $n > m + w$, and defining Y_1, Y_2 and Y_3 to be indicators associated with the new split will produce the upper bound:

$$P(X_2 = 1, X_3 = 1) \le P(Y_2 = 1) + \{P(Y_1 = 1)\}^2 \le P(Y_2 = 1) + \{P(X_1 = 1)\}^2 .$$

The probability $P(Y_2 = 1)$ is proportional to $2we^{-x}$. Consequently, b_2 is asymptotically bounded by $2\lambda[2w/m + P(X_1 = 1)]$ and it converges to zero when the threshold diverges to infinity.

The conclusion is an expression for the asymptotic distribution of $\hat{N}_x$:

$$\lim_{x \to \infty} P(\hat{N}_x \leq ye^x) = \lim_{x \to \infty} P(\hat{W} > 0) = 1 - e^{-y\{E[\mathcal{M}/\mathcal{S}]\}^2[2\Phi(CI^{\frac{3}{2}}\sigma^{-1})-1]/I} \ .$$

An alternative statement of this result is the statement that $e^{-x}\hat{N}_x$ has an asymptotic exponential distribution with a rate parameter given by:

$$\{E[\mathcal{M}/\mathcal{S}]\}^2[2\Phi(CI^{\frac{3}{2}}\sigma^{-1}) - 1]/I \ .$$

Our main goal is to obtain the asymptotic distribution of the cusum unrestricted stopping time N_x. For that we show that when C is large enough then the distribution of N_x and $\hat{N}_x$ almost coincide.

Fixing y, we want to compare the probabilities $P(N_x \leq ye^x)$ and $P(\hat{N}_x \leq ye^x)$ to each other. Luckily, one direction is automatically taken care of. By the definitions we have that $q_n \geq \hat{q}_n$, since the latter involves maximization over a smaller subset. Consequently, the process q_n must cross the threshold before (or at the same time that) the process $\hat{q}_n$ does. Hence, $N_x \leq \hat{N}_x$ and we get that $P(N_x \leq ye^x) \geq P(\hat{N}_x \leq ye^x)$. What is left to do is to show that latter probability, for a large C, is not much smaller that the former probability.

Observe that:

$$P(\hat{N}_x > ye^x) \leq P(N_x > ye^x) + P(\hat{N}_x > ye^x, N_x \leq ye^x) \ .$$

The event $\{\hat{N}_x > ye^x, N_x \leq ye^x\}$, in reference to the collection of parameters $\{(k, n) : 1 \leq k \leq n \leq ye^x\}$, corresponds to the statement that there exists a log-likelihood $\ell_{k,n}$ that goes above the threshold x, yet any such large-valued log-likelihood $\ell_{k,n}$ does not belong to the strip T that is used in the definition of $\hat{N}_x$. One may conclude that the event in question is a subset of:

$$\bigcup_{k=1}^{ye^x}\left(\left\{\max_{k \leq n < k+I/x-C\sqrt{x}} \ell_{k,n} \geq x\right\} \cup \left\{\max_{k+I/x+C\sqrt{x} < n} \ell_{k,n} \geq x\right\}\right) \ .$$

The probabilities of the event in the large brackets can be compared with probabilities defined for the sequential probability ratio test (i.e., the case $k = 1$). Each collection of events enclosed by the large brackets are of the same structure, albeit they are defined with respect to processes that initiate at time k. Probabilities are computed with respect to the null distribution that assigns the same density to all the observations. Since this process is stationary, the initiating time of the process does not matter and the probabilities of the unions within the large brackets are equal to each other for all k.

Using the exact same bounds that were used in the localization argument for the sequential probability ratio test, namely the Markov inequality applied to the log-likelihood associated with the case $n - k + 1 = x/I + C\sqrt{x}$ in order to bound the upper part of the tail and using Kolmogorov's maximal inequality to bound the lower part of the tail, will produce the overall bound:

$$P\left(\hat{N}_x > ye^x, N_x \le ye^x\right) \le \sum_{k=1}^{ye^x}\left(\frac{\sigma^2(x/I - C\sqrt{x})}{e^x C^2 I^2 x} + \frac{(x/I + C\sqrt{x})\sigma^2}{e^x C^2 I^2 x}\right) = \frac{2y\sigma^2}{C^2 I^3}.$$

The bound converges to zero when $C \to \infty$ since y is fixed.

This concludes the proof of the approximation:

$$\lim_{x \to \infty} P(N_x \le ye^x) = 1 - e^{-y\{E[\mathcal{M}/\mathcal{S}]\}^2/I}, \tag{4.5}$$

that generalizes (4.4) and validates the limit exponential distribution of $e^{-x}N_x$, with a rate given by $\{E[\mathcal{M}/\mathcal{S}]\}^2/I$. This rate results from letting $C \to \infty$.

Stopping times that are used for change-point detection in sequential quality control situations are finite with probability 1. Namely, they are eventually activated even if the production process is in control throughout. Such activation is considered a false detection and is to be delayed as much as possible.

The principle characteristic of the stopping time which is used in order to bound from below the delay in false detection is the expectation. This expectation is called the average run length to false alarm. In particular, there is interest in the asymptotic evaluation of the average run length to false alarm of the cusum change-point detection rule N_x. Knowing the relation between the average run length to detection and the threshold x allows setting the threshold in order to obey a specified average run length to false alarm that is required.

We obtained in (4.5) the asymptotic distribution of the stopping rule. That distribution was the exponential distribution with the given rate. The natural conjecture is that the limit of expectations is equal to the expectation of the limit distribution. In other words, since the expectation of an exponential random variable is the reciprocal of the rate, the conjecture suggests that:

$$\lim_{x \to \infty} e^{-x}E(N_x) = I/\{E[\mathcal{M}/\mathcal{S}]\}^2. \tag{4.6}$$

This conjecture is correct but it requires a proof.

A proof is obtained by showing that the collection of random variables $\{e^{-x}N_x\}$, indexed by x, is uniformly integrable. It is sufficient to show that the collection $\{e^{-x}\hat{N}_x\}$ is uniformly integrable for some C, since the latter collection dominates the former.

Fix C and consider the collection $\{e^{-x}\hat{N}_x\}$ as a function of x. All random variables involved are non-negative. Therefore, in order to establish uniform integrability it is sufficient to give a uniform bound on the positive tail of the distributions of the random variables, for all random variables associated with a large enough x.

Let us consider again the sequence of indicators $\{X_i\}$ that are associated with exceeding the threshold by likelihood ratio in a strip of length m and width of

$2C\sqrt{x}$ and consider a random variable U that examines every other interval and indicates the first of these intervals where a crossing of the threshold occurs. Formally, we define U via:

$$U = \inf\{k : X_{2k} = 1\},$$

where X_{2k} is the indicator of crossing the threshold in the even index $2k$.

The random variables $\{X_{2k} : k = 1, 2, \ldots\}$ are independent of each other since none share the same region of dependence. As a result, the distribution of U is geometric. This geometric distribution counts the number of even indicators that elapsed until an even indicator obtained the value 1. The probability of 'success' in this sequence of independent experiments is given by $P(X_2 = 1)$ and is asymptotic to me^{-x}.

As a result, since $\hat{N}_x \le 2mU$ we get that

$$P(e^{-x}N_x > y) \le P(U > ye^x/(2m)) = \{1 - P(X_2 = 1)\}^{ye^x/(2m)}.$$

The sequence $(e^x/m)P(X_2 = 1)$ converges to a positive limit. Thus, for some positive number δ the quantity is larger than δ for all $x > x_\delta$. Consequently,

$$P(e^{-x}N_x > y) \le \{1 - \delta me^{-x}\}^{[e^x/(m\delta)][\delta y/2]} \le e^{-\delta y/2},$$

which completes the proof of uniform integrability and establishes the approximation in (4.6).

5

The localization theorem

5.1 Introduction

In this chapter we present proofs of technical elements that are used in the derivation of the approximation of the distribution of extremes in a random field. The element of the analysis that is specific to the problem at hand is a limit argument that expresses the asymptotic value of expectations that emerge from measure transformation. The argument is summarized in a theorem that we denote the *localization theorem*. The full formulation is given in Theorem 5.2. A simplified version, where the main concepts are easier to grasp, is formulated in Theorem 5.1.

The localization theorem applies a special version of a local limit theorem for distributions. This limit theorem deals with multivariate distributions. Specifically, the joint distribution of the global term and the local field. Unlike standard formulations of limit theorems in distribution, in this version there is a distinction between the limit requirements in one direction, the direction of the global term, and the requirements for the other directions that are specified by the local field. The global term is required to obey a rate of convergence that is consistent with a local limit whereas the demand for the local field is the parallel of standard convergence in distribution.

In general, the limit of the local field need not be Gaussian. However, if the local field is a sum of independent fields it will typically be Gaussian in the limit. In several applications that we will encounter the global term and the local field are produced from the summation of independent and identically distributed fields. In Theorem 5.3 we show that the minimal conditions of a one-dimensional local limit theorem and a finite-dimension central limit theorem, when combined with a vanishing correlation between the global term and the local field, is suffi-cient in order to establish the local limit theorem that is required by Theorem 5.2.

Extremes in Random Fields: A Theory and its Applications, First Edition. Benjamin Yakir.

5.2   A simplified version of the localization theorem

Consider a triangular array, in which κ is the primary index. Let $\tilde{\ell}_\kappa$ be a random variable that is closely related to the global log-likelihood ratio. Typically, it will be the difference between the log-likelihood ratio and a threshold. This difference is expected to converge after rescaling (dividing, say, by $\kappa^{\frac{1}{2}}$) to a normal distribution with a possibly non zero mean. Let M_κ and S_κ be a pair of random variables, measurable with respect to the collection of observations. Typically, $\log M_\kappa$ will correspond to the maximization of some random field, and S_κ will be the sum of the exponentiated random field. In many cases M_κ will be completely analogous to S_κ, but with the summation replaced by maximization. In such cases the ratio M_κ / S_κ will have values between zero and one.

After a change of measure along the lines discussed in preceding chapters, we are interested in the limit, as $\kappa \to \infty$, of the term

$$\kappa^{\frac{1}{2}} \mathrm{E}[(M_\kappa / S_\kappa) \exp[-(\tilde{\ell}_\kappa + \log M_\kappa)]; \tilde{\ell}_\kappa + \log M_\kappa \geq 0] .$$

We first apply a localization step that replaces the quantities M_κ and S_κ by local versions that are almost independent of $\tilde{\ell}_\kappa$. A local central limit theorem is applied to the conditional distribution of $\tilde{\ell}_\kappa$, and distributional approximations are applied to the localized version of M_κ / S_κ. Together these yield the desired limit.

Consider a sequence of σ-fields $\hat{\mathcal{F}}_\kappa, \kappa = 1, 2, \ldots$. Let $\hat{M}_\kappa$ and $\hat{S}_\kappa$ be approximations of M_κ and S_κ, respectively, which are measurable with respect to $\hat{\mathcal{F}}_\kappa$. Given $\epsilon > 0$, we will assume that for all large κ:

I. M_κ and S_κ satisfy $0 < M_\kappa \leq S_\kappa < \infty$ with probability one.

II. There exist $\hat{M}_\kappa$ and $\hat{S}_\kappa$ measurable with respect to $\hat{\mathcal{F}}_\kappa$ such that $|\hat{M}_\kappa / M_\kappa - 1| \leq \epsilon$ and $|\hat{S}_\kappa / S_\kappa - 1| \leq \epsilon$, with probability at least $1 - \epsilon \kappa^{-\frac{1}{2}}$.

III. $\mathrm{E}[\hat{M}_\kappa / \hat{S}_\kappa]$ converges to a finite and positive limit denoted by $\mathrm{E}[\mathcal{M}/\mathcal{S}]$.

IV. There exist $\mu_\kappa \in \mathbb{R}$ and $\sigma_\kappa \in \mathbb{R}^+$ such that for every $0 < \epsilon, \delta$ and for all large enough κ the probability of the event

$$B_\kappa = \left\{ \sup_{|x| \leq \log \kappa} \left| \kappa^{\frac{1}{2}} \mathrm{P}\big(\tilde{\ell}_\kappa \in x - \log \hat{M}_\kappa + (0, \delta] | \hat{\mathcal{F}}_\kappa\big) - \frac{\delta}{\sigma_\kappa} \phi\left(\frac{\mu_\kappa}{\sigma_\kappa}\right) \right| \leq \epsilon \right\}$$

is bounded from below by $1 - \epsilon \kappa^{-\frac{1}{2}}$.

Theorem 5.1   *If Conditions I–IV hold and $\mu_\kappa \to \mu$, $\sigma_\kappa \to \sigma > 0$, then*

$$\lim_{\kappa \to \infty} \kappa^{\frac{1}{2}} \mathrm{E}\left[(M_\kappa / S_\kappa) e^{-(\tilde{\ell}_\kappa + \log M_\kappa)}; \tilde{\ell}_\kappa + \log M_\kappa \geq 0\right] = \sigma^{-1} \phi(\mu/\sigma) \mathrm{E}[\mathcal{M}/\mathcal{S}] ,$$

where ϕ is the density of the standard normal distribution.

Before giving a proof of the theorem, we consider the conditions. The random variable M_κ typically corresponds to the maximization of a finite collection of non-negative random variables, and S_κ corresponds to summation over the same collection. Condition I is satisfied in such a case. The σ-field $\hat{\mathcal{F}}_\kappa$ contains the essential information for the determination of the local behavior of the likelihood ratios. Conditions II and III deal with approximation of the functionals that summarize the effect of the local process by the use of terms that are functions of the local field.

The global component $\tilde{\ell}_\kappa$ is approximately normal. Condition IV provides a local central limit statement and establishes a suitable relation between the local components and the global one.

Proof of Theorem 5.1.   Denote by A_κ the event over which Condition II holds. We will produce both an upper bound and a lower bound, which together prove the theorem.

Start with an upper bound. On the event A_κ we replace S_κ by the lower bound $\hat{S}_\kappa/(1 + \epsilon)$ and replace M_κ by an upper bound given in terms of $\hat{M}_\kappa$. Specifically, we replace $\log M_\kappa$ on the event $\{\tilde{\ell}_\kappa + \log(M_\kappa) \geq 0\}$ by the upper bound $\log \hat{M}_\kappa - \log(1 - \epsilon)$ in order to obtain:

$$E\left[(M_\kappa/S_\kappa)e^{-(\tilde{\ell}_\kappa + \log M_\kappa)}; \tilde{\ell}_\kappa + \log M_\kappa \geq 0\right] \leq P(A_\kappa^c)$$

$$+ \frac{1+\epsilon}{1-\epsilon}E\left[(\hat{M}_\kappa/\hat{S}_\kappa)e^{-(\tilde{\ell}_\kappa + \log \hat{M}_\kappa - \log(1-\epsilon))}; \tilde{\ell}_\kappa\right.$$

$$\left. + \log \hat{M}_\kappa - \log(1 - \epsilon) \geq 0, A_\kappa\right].$$

In this inequality we exploit the fact that

$$0 < (M_\kappa/S_\kappa)e^{-(\tilde{\ell}_\kappa + \log M_\kappa)}I_{\{\tilde{\ell}_\kappa + \log M_\kappa\}} \leq 1\,,$$

thus replacing the expectation over the complementary of the event A_κ by the probability of the complementary event. Due to the assumption in Condition II we can ignore $P(A_\kappa^c)$.

Denote $\hat{m}_\kappa = \log \hat{M}_\kappa - \log(1 - \epsilon)$. By conditioning on the σ-field $\hat{\mathcal{F}}_\kappa$ and the fact that $\hat{M}_\kappa$ and $\hat{S}_\kappa$ are measurable with respect to $\hat{\mathcal{F}}_\kappa$ one finds that:

$$E\left[\frac{\hat{M}_\kappa}{\hat{S}_\kappa}e^{-(\tilde{\ell}_\kappa + \hat{m}_\kappa)}; \tilde{\ell}_\kappa + \hat{m}_\kappa \geq 0\right] = E\left[\frac{\hat{M}_\kappa}{\hat{S}_\kappa}E\left(e^{-(\tilde{\ell}_\kappa + \hat{m}_\kappa)}; \tilde{\ell}_\kappa + \hat{m}_\kappa \geq 0|\hat{\mathcal{F}}_\kappa\right)\right].$$

We proceed the construction of an upper bound by analyzing the conditional expectation. The exponential function is a monotone function. Hence:

$$E\left(e^{-(\tilde{\ell}_\kappa + \hat{m}_\kappa)}; \tilde{\ell}_\kappa + \hat{m}_\kappa \geq 0|\hat{\mathcal{F}}_\kappa\right) \leq$$

$$\sum_{j=0}^{\lceil \log(\kappa)/\delta \rceil} e^{-\delta j}P\left(\tilde{\ell}_\kappa \in \delta j - \hat{m}_\kappa + (0, \delta]|\hat{\mathcal{F}}_\kappa\right) + \frac{1}{\kappa(1 - e^{-\delta})}\,.$$

After multiplication by $\kappa^{\frac{1}{2}}$ we get, directly from Condition IV, that over the event B_κ the sum is bounded from above by

$$\left[\delta/(1 - e^{-\delta})\right]\left[\sigma_\kappa^{-1}\phi(\mu_\kappa/\sigma_\kappa) + \epsilon\right]$$

and the remainder term still converges to zero. Taking expectation with respect to $\hat{\mathcal{F}}_\kappa$ and using Condition III and the fact that the ratio $\hat{M}_\kappa/\hat{S}_\kappa$ is bounded, eliminating in the process the vanishingly small error caused by the complementary of the event B_κ, complete the upper bound.

The lower bound is constructed in a similar way. This time we replace S_κ by an upper bound and replace M_κ by a lower bound. This produces:

$$\mathrm{E}\left[(M_\kappa/S_\kappa)e^{-(\tilde{\ell}_\kappa + \log M_\kappa)}; \tilde{\ell}_\kappa + \log M_\kappa \geq 0\right] \geq$$

$$\frac{1-\epsilon}{1+\epsilon}\mathrm{E}\left[(\hat{M}_\kappa/\hat{S}_\kappa)e^{-(\tilde{\ell}_\kappa + \log \hat{M}_\kappa - \log(1+\epsilon))}; \tilde{\ell}_\kappa + \log \hat{M}_\kappa - \log(1+\epsilon) \geq 0, A_\kappa\right].$$

Denote now $\hat{m}_\kappa = \log \hat{M}_\kappa - \log(1 + \epsilon)$ and consider the expectation on the right-hand side:

$$\mathrm{E}\left[(\hat{M}_\kappa/\hat{S}_\kappa)e^{-(\tilde{\ell}_\kappa + \hat{m}_\kappa)}; \tilde{\ell}_\kappa + \hat{m}_\kappa \geq 0, A_\kappa\right] \geq$$

$$\mathrm{E}\left[(\hat{M}_\kappa/\hat{S}_\kappa)e^{-(\tilde{\ell}_\kappa + \hat{m}_\kappa)}; \tilde{\ell}_\kappa + \hat{m}_\kappa \geq 0, B_\kappa\right] - \mathrm{P}(A_\kappa^c).$$

As before, we will ignore the error resulting from the complementary of the event A_κ and use the monotonicity of the exponential function:

$$\mathrm{E}\left(e^{-(\tilde{\ell}_\kappa + \hat{m}_\kappa)}; \tilde{\ell}_\kappa + \hat{m}_\kappa \geq 0 | \hat{\mathcal{F}}_\kappa\right) \geq$$

$$\sum_{j=0}^{\lfloor \log(\kappa)/\delta \rfloor - 1} e^{-\delta(j+1)}\mathrm{P}\left(\tilde{\ell}_\kappa \in \delta j - \hat{m}_\kappa + (0, \delta] | \hat{\mathcal{F}}_\kappa\right).$$

Observe that we are using the fact that B_κ is measurable with respect to $\hat{\mathcal{F}}_\kappa$.

After multiplication by $\kappa^{\frac{1}{2}}$ we get, directly from Condition IV, that over the event B_κ the sum is bounded from below by:

$$e^{-\delta}(1 - 1/\kappa)\left[\delta/(1 - e^{-\delta})\right]\left[\sigma_\kappa^{-1}\phi(\mu_\kappa/\sigma_\kappa) - \epsilon\right].$$

Taking expectation with respect to $\hat{\mathcal{F}}_\kappa$, using again Condition III and the fact that the ratio $\hat{M}_\kappa/\hat{S}_\kappa$ is bounded, complete the lower bound and the proof of Theorem 5.1. □

5.3   The localization theorem

The conditions for Theorem 5.1 produce a simple proof. In simple cases, for example when the random field is Gaussian to begin with, one may be able to verify the conditions. However, in more complex scenarios the verification is more difficult. In particular, the most problematic condition is Condition IV, which is a form of a conditional local central limit theorem of the global term, given the σ-algebra generated by the local field.

There are two difficulties with the condition as stated. First, the localization argument that is used in the proof of Theorem 5.1 requires that the approximation of M_κ and S_κ by $\hat{M}_\kappa$, respectively $\hat{S}_\kappa$, improves as κ increases. This, in turn, calls for an increase in the number of elements of the local field that are involved in the approximation. Consequently, the consideration of a conditional local central limit theorem for $\tilde{\ell}_\kappa$, given only a bounded number of elements of the local field, will not be sufficient.

The other difficulty is that a conditional local central limit theorem is stronger than what is required. Actually, considering the joint distribution of the global term and the local field, only the part that involves the global term demands a local limit theorem. For the part that involves the local field a weak convergence is sufficient. Consideration of conditional distribution essentially calls for the investigation of ratios of densities, thereby enforcing an approximation of the density in the denominator, namely the density of the local field.

As a remedy for the shortcomings of Theorem 5.1 we propose a modified version of the theorem. The conditions in this modified version are more difficult to state but are easier to validate. The price that we pay is in the form of a more complex proof.

Given $\epsilon, \epsilon_3 > 0$, we assume that for some function $g(\kappa)$, $\log \kappa \le g(\kappa) \le \epsilon \kappa^{\frac{1}{2}}$, for some $C < \infty$, and for all large κ the following conditions:

I*: M_κ, S_κ, $\hat{M}_\kappa$ and $\hat{S}_\kappa$ satisfy $0 \le M_\kappa/S_\kappa \le C$ and $0 \le \hat{M}_\kappa/\hat{S}_\kappa \le C$ with probability one.

II*: Denote $A_{\mathrm{II}}^c = \{|\log M_\kappa - \log \hat{M}_\kappa| > \epsilon\} \cup \{|\hat{S}_\kappa/S_\kappa - 1| > \epsilon\}$. For some $0 < \delta$ that does not depend on ϵ:

$$\max_{|x| \le 3g(\kappa)} P(A_{\mathrm{II}}^c \cap \{\tilde{\ell}_\kappa + \log \hat{M}_\kappa \in x + (0, \delta]\} \cap \{|\hat{m}| \le g(\kappa)\}) \le \epsilon \kappa^{-\frac{1}{2}}.$$

(5.1)

III*: $E[\hat{M}_\kappa/\hat{S}_\kappa]$ converges to $E[\hat{\mathcal{M}}/\hat{\mathcal{S}}]$ and $|E[\hat{\mathcal{M}}/\hat{\mathcal{S}}] - E[\mathcal{M}/\mathcal{S}]| \le \epsilon_3$.

IV*: There exist $\mu \in \mathbb{R}$ and $\sigma \in \mathbb{R}^+$ such that for every $0 < \epsilon_4, \delta$, for any event $E \in \hat{\mathcal{F}}_\kappa$ having boundary measure 0, and for all large enough κ:

$$\sup_{|x| \le 3g(\kappa)} \left| \kappa^{\frac{1}{2}} P(\tilde{\ell}_\kappa + \log \hat{M}_\kappa \in x + (0, \delta], E) - \frac{\delta}{\sigma} \phi\left(\frac{\mu}{\sigma}\right) P(E) \right| \le \epsilon_4,$$

and also:

V*: $P(|\log M_\kappa| > g(\kappa))$, $P(|\log \hat{M}_\kappa| > g(\kappa))$ and $P(\log M_\kappa - \log \hat{M}_\kappa < -\epsilon)$ are all $o(\kappa^{-\frac{1}{2}})$.

Theorem 5.2   *If Conditions I^*-V^* hold then:*

$$\lim_{\kappa \to \infty} \kappa^{\frac{1}{2}} E[(M_\kappa/S_\kappa)e^{-(\tilde{\ell}_\kappa + \log M_\kappa)}; \tilde{\ell}_\kappa + \log M_\kappa \geq 0] = \sigma^{-1}\phi(\mu/\sigma)E[\mathcal{M}/\mathcal{S}].$$

Conditions I^* and III^* are essentially the same as Conditions I and III of Theorem 5.1. Condition II^* is less restrictive than the parallel condition of Theorem 5.1. In the previous condition the probability of the event where the approximations of the maximum and the sum is not valid was required to go to zero at a rate faster than $\kappa^{-\frac{1}{2}}$. In the new condition that rate should hold when intersecting the event with an event that already has a rate of convergence of $O(\kappa^{-\frac{1}{2}})$. Condition IV^* is the statement of the local limit theorem. The condition is formulated in terms of the joint distribution of the local field and $\tilde{\ell}_\kappa$ and is substantially weaker than the corresponding requirement in Condition IV, which involves conditional distribution of the latter given the former. Condition V^* is a new condition. It is a natural condition when localization involves both a reduction in the set of indices that enter into the maximization/summation and an approximation of the local process. Note that the sum and the maximum of likelihood ratios are both decreasing if the set of indices is reduced. Hence, the actual requirement is for an improving approximation of the elements of the local field.

Notice that there is another discrepancy between Condition IV and Condition IV^*. In the former the maximization of the error is taken with respect to x in the region $|x| \leq \log \kappa$. In the latter the maximization is over a larger region $|x| \leq g(\kappa)$. There is a tradeoff between Conditions II^* and IV^* and Condition V^*. The better we can control the magnitude of the local field the less stringent the bound in the local limit theorem that we need to validate for the global term.

Proof of Theorem 5.2   The proof is similar to that of Theorem 5.1, but the details are more involved.

Consider first an upper bound. Denote by A_V^c the union of the events entering into Condition V^*. As before, on A_{II}, the complementary of A_{II}^c from Condition II^*, we decrease S_κ and increase $\log M_\kappa$ in the event $\{\tilde{\ell}_\kappa + \log \hat{M}_\kappa \geq 0\}$ to obtain:

$$E\left[(M_\kappa/S_\kappa)e^{-(\tilde{\ell}_\kappa + \log M_\kappa)}; \tilde{\ell}_\kappa + \log M_\kappa \geq 0\right] \leq$$

$$\frac{1+\epsilon}{1-\epsilon}E\left[(\hat{M}_\kappa/\hat{S}_\kappa)e^{-(\tilde{\ell}_\kappa + \log \hat{M}_\kappa - \log(1-\epsilon))}; \tilde{\ell}_\kappa + \log \hat{M}_\kappa - \log(1-\epsilon) \geq 0, A_V\right]$$

$$+ CE\left[e^{-(\tilde{\ell}_\kappa + \log M_\kappa)}; \tilde{\ell}_\kappa + \log M_\kappa \geq 0, A_{II}^c \cap A_V\right] + CP(A_V^c).$$

We used the fact that $M_\kappa/S_\kappa \leq C$ in the last two terms. By Conditions I^* and V^* we can ignore the last of these terms.

Denote $\hat{m}_\kappa = \min\{\log \hat{M}_\kappa, g(\kappa)\} - \log(1-\epsilon)$ and consider the first expectation. Notice that $\hat{M}_\kappa/\hat{S}_\kappa$ is a bounded function of the local process. Let $\sum_{u=1}^n e_u I_{E_u}$ be a simple function that bounds the function from above and approximates it up to ϵ in the maximal norm. The partition $\{E_u\}$ may be formed with sets of zero boundary that belong to $\hat{\mathcal{F}}_\kappa$. The number of events in the partition $n = n(\epsilon)$ is finite for any positive ϵ. By the monotonicity of the exponential function:

$$E\left[(\hat{M}_\kappa/\hat{S}_\kappa)e^{-(\tilde{\ell}_\kappa + \hat{m}_\kappa)}; \, \tilde{\ell}_\kappa + \hat{m}_\kappa \geq 0\right] \leq$$

$$\sum_{j=0}^{\lceil \log(\kappa)/\delta \rceil} \sum_u e_u e^{-j\delta} P\left(\tilde{\ell}_\kappa + \hat{m}_\kappa \in j\delta + (0, \delta], E_u\right) + \frac{1}{\kappa(1 - e^{-\delta})} \, .$$

After multiplication by $\kappa^{\frac{1}{2}}$ we find by Conditions IV* and V* that the sum is bounded over the event defined in these conditions by:

$$\frac{\delta}{1 - e^{-\delta}} \frac{1}{\sigma} \phi\left(\frac{\mu}{\sigma}\right) \sum_u e_u P(E_u) + \frac{\epsilon_4}{1 - e^{-\delta}} \sum_u e_u \, .$$

Since $\sum_u e_u P(E_u)$ approximates the expectation of $\hat{M}_\kappa/\hat{S}_\kappa$ and since $\sum_u e_u$ is bounded and ϵ_4 can be made arbitrarily small relative to both ϵ and δ, we get the upper bound associated with the first expectation.

In order to complete the upper bound we need to show that

$$\kappa^{\frac{1}{2}} E\left[e^{-(\tilde{\ell}_\kappa + \log M_\kappa)}; \, \tilde{\ell}_\kappa + \log M_\kappa \geq 0, A_{\mathrm{II}}^c \cap A_{\mathrm{V}}\right] \tag{5.2}$$

is negligible.

We examine the expectation over each of the events $\{|\log M_\kappa| > g(\kappa)\}$, $\{|\log M_\kappa| \leq g(\kappa), |\tilde{\ell}_\kappa| > 2g(\kappa)\}$, and $\{|\tilde{\ell}_\kappa| \leq 2g(\kappa)\}$. The expectation over the first event is negligible by Condition V*. The expectation over the second event is negligible because the event implies that $\tilde{\ell}_\kappa + \log(M_\kappa) > g(\kappa)$. Consequently, we get a vanishingly small bound of order $\kappa^{\frac{1}{2}} \exp\{-g(\kappa)\}$. Hence, we only need to deal with the expectation over the third event.

Let $m_\kappa = \log M_\kappa$, $\hat{m}_\kappa = \log \hat{M}_\kappa$. Set $B_0 = \{|m_\kappa - \hat{m}_\kappa| < \epsilon\}$, and for $0 < i \leq 2g(\kappa)/\delta$, let $B_i = \{m_\kappa - \hat{m}_\kappa \in \epsilon + (i - 1)\delta + (0, \delta]\}$. Observe that $A_{\mathrm{II}}^c \cap A_{\mathrm{V}} \subset \cup_{i>0}(B_i \cap A_{\mathrm{II}}^c) \subset A_{\mathrm{II}}^c$.

Let $C_j = \{\tilde{\ell}_\kappa + \hat{m}_\kappa \in j\delta + (0, \delta]\} \cap \{|\hat{m}_\kappa| \leq g(\kappa)\}$. This event is a function of $\tilde{\ell}_\kappa$ and the local σ-algebra $\hat{\mathcal{F}}_\kappa$. Conditioning on $\hat{\mathcal{F}}_\kappa$ and on $\tilde{\ell}_\kappa$ we get

$$E\left[e^{-(\tilde{\ell}_\kappa + m_\kappa)}; \, \tilde{\ell}_\kappa + m_\kappa \geq 0, A_{\mathrm{II}}^c \cap A_{\mathrm{V}} \cap B_i \cap C_{j-i} \Big| \hat{\mathcal{F}}_\kappa, \tilde{\ell}_\kappa\right]$$

$$\leq e^\delta e^{-j\delta} I_{\{j \geq 0\}} I_{C_{j-i}} P\left(A_{\mathrm{II}}^c \cap B_i, \big| \hat{\mathcal{F}}_\kappa, \tilde{\ell}_\kappa\right)$$

$$\leq e^\delta e^{-j\delta} I_{\{j \geq 0\}} I_{C_{j-i}} P\left(A_{\mathrm{II}}^c \big| \hat{\mathcal{F}}_\kappa, \tilde{\ell}_\kappa\right) \, .$$

The value of $\tilde{\ell}_\kappa + m_\kappa$ is essentially equal to the quantity $j\delta$. The largest value of j that needs to be considered is $j = 3g(\kappa)/\delta$. The smallest value is $j = 0$. The fact that only $j \geq 0$ needs to be considered follows from the fact that $\tilde{\ell}_\kappa + m_\kappa \geq 0$.

Taking expectations with respect to the conditional distribution, given $\hat{\mathcal{F}}_\kappa$, produces:

$$E\left[e^{-(\tilde{\ell}_\kappa + m_\kappa)}; \tilde{\ell}_\kappa + m_\kappa \geq 0, A_{\mathrm{II}}^c \cap A_{\mathrm{V}}, |\tilde{\ell}_\kappa| \leq 2g(\kappa) \big| \hat{\mathcal{F}}_\kappa \right]$$

$$\leq e^\delta \sum_{j=0}^{\lceil 3g(\kappa)/\delta \rceil} \sum_{i=0}^{j} e^{-j\delta} E\left[P\left(A_{\mathrm{II}}^c \big| \hat{\mathcal{F}}_\kappa, \tilde{\ell}_\kappa\right); C_{j-i} \big| \hat{\mathcal{F}}_\kappa \right]$$

$$= e^\delta \sum_{j=0}^{\lceil 3g(\kappa)/\delta \rceil} \sum_{i=0}^{j} e^{-j\delta} P\left(A_{\mathrm{II}}^c \cap \{\tilde{\ell}_\kappa + \hat{m}_\kappa \in (j-i)\delta + (0, \delta]\} \big| \hat{\mathcal{F}}_\kappa\right).$$

When we integrate with respect to the local random field and over the event $D = \{|\hat{m}| \leq g(\kappa)\}$, we get:

$$P(A_{\mathrm{II}}^c \cap \{\tilde{\ell}_\kappa + \hat{m}_\kappa \in (j-i)\delta + (0, \delta]\} \cap D)$$

$$\leq \max_{|x| \leq 3g(\kappa)} P(A_{\mathrm{II}}^c \cap \{\tilde{\ell}_\kappa + \hat{m}_\kappa \in x + (0, \delta]\} \cap D)$$

that does not depend any longer on i or j. Multiplying by $\kappa^{\frac{1}{2}}$ and applying Condition II* leads to the bound:

$$\frac{e^\delta(1 + \epsilon)}{(1 - e^{-\delta})^2} \times \epsilon.$$

This completes the proof of the upper bound.

Now we consider the lower bound. This time we increase S_κ and decrease M_κ in the event $\{\tilde{\ell}_\kappa + \log \hat{M}_\kappa \geq 0\}$ to obtain

$$E\left[\frac{M_\kappa}{S_\kappa} e^{-(\tilde{\ell}_\kappa + \log M_\kappa)}; \tilde{\ell}_\kappa + \log M_\kappa \geq 0\right] \geq$$

$$\frac{1-\epsilon}{1+\epsilon} E\left[\frac{\hat{M}_\kappa}{\hat{S}_\kappa} e^{-(\tilde{\ell}_\kappa + \log \hat{M}_\kappa - \log(1+\epsilon))}; \tilde{\ell}_\kappa + \log \hat{M}_\kappa - \log(1+\epsilon) \geq 0, A_{\mathrm{II}} \cap A_{\mathrm{V}}\right].$$

Denote now $\hat{m}_\kappa = \log \hat{M}_\kappa - \log(1 + \epsilon)$ and consider the expectation on the right-hand side:

$$E[(\hat{M}_\kappa/\hat{S}_\kappa)e^{-(\tilde{\ell}_\kappa + \hat{m}_\kappa)}; \tilde{\ell}_\kappa + \hat{m}_\kappa \geq 0, A_{\mathrm{II}} \cap A_{\mathrm{V}}] =$$

$$E[(\hat{M}_\kappa/\hat{S}_\kappa)e^{-(\tilde{\ell}_\kappa + \hat{m}_\kappa)}; \tilde{\ell}_\kappa + \hat{m}_\kappa \geq 0]$$

$$- E[(\hat{M}_\kappa/\hat{S}_\kappa)e^{-(\tilde{\ell}_\kappa + \hat{m}_\kappa)}; \tilde{\ell}_\kappa + \hat{m}_\kappa \geq 0, A_{\mathrm{II}}^c, A_{\mathrm{V}}] - CP(A_{\mathrm{V}}^c).$$

We may ignore the last probability. For the first term on the right-hand side of the equation we get by the monotonicity of the exponential function:

$$E[(\hat{M}_\kappa/\hat{S}_\kappa)e^{-(\tilde{\ell}_\kappa+\hat{m}_\kappa)}; \tilde{\ell}_\kappa + \hat{m}_\kappa \geq 0] \geq$$

$$\sum_{j=0}^{\lceil \log(\kappa)/\delta \rceil} \sum_u e_u e^{-j\delta} P(\tilde{\ell}_\kappa + \hat{m}_\kappa \in j\delta + (0, \delta], E_u) I_{\{\hat{m}_\kappa < g(\kappa)\}},$$

where this time $\sum_u e_u E_u$ is a simple and non-negative function that approximates $\hat{M}_\kappa/\hat{S}_\kappa$ from below. After multiplication by $\kappa^{\frac{1}{2}}$ and the application of Condition IV* we get (over the event $\{\hat{m}_\kappa < g(\kappa)\}$) the lower bound:

$$\frac{e^{-\delta}\delta(1 - 1/\kappa)}{1 - e^{-\delta}} \frac{1}{\sigma} \phi\left(\frac{\mu}{\sigma}\right) \sum_u e_u P(E_u) - \frac{\epsilon_4}{1 - e^{-\delta}} \sum_u e_u .$$

Again, ignoring the negligible reduction that is caused by the truncation $\{\hat{m}_\kappa < g(\kappa)\}$, we get a first term which is almost equal to the leading term that appears in the upper bound.

In order to complete the proof we need an upper bound on the term

$$E[(\hat{M}_\kappa/\hat{S}_\kappa)e^{-(\tilde{\ell}_\kappa+\hat{m}_\kappa)}; \tilde{\ell}_\kappa + \hat{m}_\kappa \geq 0, A_{II}^c, A_V] .$$

But this can be obtained by a repeat of the proof that was used in order to bound (5.2), the parallel term that emerged in the context of constructing an upper bound.   □

5.4   A local limit theorem

Condition IV* of Theorem 5.2 is an interesting condition and it is formulated the way it is for a reason. The context of the condition are local limit theorems and higher order approximation in the central limit theorem. The two issues are related. Consider a distribution that converges to the normal distribution. If an expansion with an error of order $o(n^{-\frac{1}{2}})$ is given to the distribution function then that expansion can be used in order to produce a local limit type of approximation.

In the one-dimensional independently and identically distributed (i.i.d.) setting such an approximation exists under very mild conditions. Evidence is Theorem A.1, which requires only a second moment and the minimal assumption of a non lattice distribution. Adding a third moment will guarantee a uniform approximation.

However, in the multivariate case the picture is not as simple (See for example [10]). An attempt to obtain an accurate approximation of the joint distribution is not trivial at all and typically requires conditions that some applications do not share. The best result we know of that may be used in our context is Theorem 20.1 in [11] that produces an expansion of the distribution of a standardized sum of i.i.d. random vectors with an accuracy that depends on the highest moment. In particular, if a third moment exists then the expansion seems to be good enough for our needs.

There is a catch. The joint characteristic function in the theorem is required to satisfy the Cramér condition. This condition states that the absolute value of the characteristic function must not converge to one along any subsequence that diverges to infinity. In comparison, the non lattice assumption in the one-dimensional case simply states that the value is one only at the origin. The non lattice condition does not exclude cases where the absolute value converges to one for a sequence that converges to infinity. The Cramér condition does not allow such convergence.

This condition would not have been a serious concern. Alas, it excludes discrete distributions, preventing us from using for example functions of the binomial or Poisson distributions. Consequently, it is worthwhile to look for proofs that are less restrictive.

An important point in Condition IV* is the difference between the global term, which requires an accurate assessment of the distribution, and the local field. For the local field, the multivariate part, only weak convergence is called for. The rate of that convergence is not critical. Consequently, one may hope that the situation is closer to the situation encountered in the one-dimensional case and more general proofs with less restrictive conditions can be obtained.

Theorem 5.3 is an attempt in that direction. It deals with the situation when the field is a sum of i.i.d. finite dimensional fields. One coordinate in each of the fields, the coordinate associated with the global term, has variance 1. The other coordinates share a vanishing covariance structure that converges to a limit when summed. The correlation between the (standardized) global term and the local field is vanishing in the limit, which implies independence between the limit Gaussian local field and the global term. It should be noted that this theorem does not cover situations, such as the case of the sequential probability ratio rest, where the limit of the local field is not Gaussian.

Theorem 5.3   *Let $X_1, X_2, \ldots, X_n$ be i.i.d. random vectors of dimension $d + 1$. Assume that $\mathrm{E}(X_1) = (0, \mu_n)'$ and*

$$\mathrm{Var}(X_1) = \begin{pmatrix} 1 & \rho'_n \\ \rho_n & \Sigma_n \end{pmatrix},$$

where ρ_n is a correlation vector and Σ_n is a d-dimensional variance-covariance matrix. Assume that the marginal distribution of the first coordinate is non-lattice and identical for all n. Consider the sum $S_n = X_1 + \cdots + X_n$ and let $m = m_n : \mathbb{R}^d \to \mathbb{R}$ be a bounded function. The term $m(S_n)$ corresponds to the application of the function m to the last d coordinates of S_n. Then if $\delta > 0$ and a set A with a zero measure boundary are fixed and if $x_n/\sqrt{n} \to x$, for a finite x, $n\mu_n \to \mu$,

$n\Sigma_n \to \Sigma$, $n^{\frac{1}{2}}\rho_n \to 0$, and $\max |m| = o(\sqrt{n})$ then

$$\lim_{n\to\infty} \sqrt{n} P(S_n + m(S_n) \in (x_n, x_n + \delta) \times A) = \delta\phi(x)P(Z \in A) ,$$

for $Z \sim N(\mu, \Sigma)$.

Proof. Let $\delta > 0$, set $S_n = X_1 + \cdots X_n$, and assume initially that $\mu_n = 0$ and $m \equiv 0$. Use a kernel produced of the joint density function of independent Polya's random variables:

$$H_0(y, z) = h(y)k(z) = \frac{1}{\pi}\frac{1 - \cos(\epsilon_0 y)}{\epsilon_0 y^2} \times \prod_{j=1}^{d}\left[\frac{1}{\pi}\frac{1 - \cos(\epsilon_j z_j)}{\epsilon_j z_j^2}\right],$$

for $\epsilon_j > 0$, $0 \le j \le d$, y a real number and $z = (z_1, \ldots, z_d)'$ a vector.

Consider the characteristic function of this Polya kernel, a product of the marginal characteristic functions, which is of a bounded support and is equal to:

$$\hat{H}_0(u_0, u) = \hat{h}(u_0)\hat{k}(u) = \begin{cases} \prod_{j=0}^{d}(1 - |u_j/\epsilon_j|) & \text{if, for all } j, |u_j| \le \epsilon_j, \\ 0 & \text{otherwise.} \end{cases}$$

Extend the density to a family of complex-valued functions by taking

$$H_\theta(y, z) = h_\vartheta(y)k_\eta(z) = e^{i\vartheta y}h(y)e^{i\langle\eta,z\rangle}k(z) = e^{(i\vartheta y + i\langle\eta,z\rangle)}H_0(y, z) ,$$

for $\theta = (\vartheta, \eta)$, and observe that $\hat{H}_\theta(u_0, u) = \hat{H}_0(u_0 + \vartheta, u + \eta)$.

We start by showing that for any given θ:

$$\lim_{n\to\infty} \sqrt{n} E H_\theta(S_n - (x_n, 0)') = \phi(x) \int h_\vartheta(y) \, dy \cdot E k_\eta(Z) , \tag{5.3}$$

where ϕ is the density of the standard normal distribution and $Z \sim N(0, \Sigma)$. For that we use the inversion formula for characteristic functions and get that:

$$H_0(y, z) = (2\pi)^{-(d+1)} \iint e^{-iu_0 y - i\langle u, z\rangle} \hat{H}_0(u_0, u) \, du_0 du$$

and therefore, by the change of variable $u_0 = v + \vartheta$, $u = w + \eta$

$$H_\theta(y, z) = e^{i\vartheta y + i\langle\eta,z\rangle} H_0(y, z)$$

$$= \frac{1}{(2\pi)^{d+1}} \iint e^{-i(u_0-\theta)y - i\langle u-\eta,z\rangle} \hat{H}_0(u_0, u) \, du_0 du$$

$$= \frac{1}{(2\pi)^{d+1}} \iint e^{-ivx - i\langle w,z\rangle} \hat{H}_\theta(v, w) \, dvdw .$$

Denote the distribution of $S_n - (x_n, 0)'$ by F_n and apply Fubini's theorem to write:

$$\mathrm{E}H_\theta(S_n - (x_n, 0)') = \frac{1}{(2\pi)^{d+1}} \iiint e^{-ivy - i\langle w, z\rangle} \hat{H}_\theta(v, w) \, dv dw dF_n(y, z)$$

$$= \frac{1}{(2\pi)^{d+1}} \iiint e^{-ivy - i\langle w, z\rangle} dF_n(y, z) \hat{H}_\theta(v, w) \, dv dw \ .$$

The innermost integral corresponds to the characteristic function of $S_n - (x_n, 0)'$, evaluated at $(-v, -w)'$, hence

$$= \frac{1}{(2\pi)^{d+1}} \iint [\varphi_n(-v, -w)]^n e^{ivx_n} \hat{H}_\theta(v, w) \, dv dw \ .$$

Notice that

$$|\varphi_n(-v, -w) - \varphi_n(-v, 0)| \le \mathrm{E}|e^{-i\langle(0, w), X_1\rangle} - 1| = o(w' \Sigma_n w) \ , \qquad (5.4)$$

so the joint characteristic function may be bounded by the marginal characteristic function of the first component with an error of order smaller than $1/n$, uniformly over bounded regions of w. Moreover, the joint characteristic function is bounded by

$$|\varphi_n(-v, -w)| \le \exp\{-(1/4)[v^2 + 2v\langle \rho_n, w\rangle + w' \Sigma_n w]\} \ , \qquad (5.5)$$

whenever the term in the square brackets is small enough.

In order to show that the limit of the given integral, multiplied by $\sqrt{n}$, is equal to the right-hand side of (5.3) we consider four regions. The first region is the region $[-\epsilon_n, \epsilon_n] \times B$, where $B = [-M, M]^d$ and where $\epsilon_n = \log(n)/n^{\frac{1}{2}}$. The second region is $([-\epsilon, -\epsilon_n) \cup (\epsilon_n, \epsilon]) \times B$, for $\epsilon > 0$ small enough to assure the validity, for all $n \ge n_\epsilon$, of the bound (5.5). The third region is $([-M, M] \times B) \setminus ([-\epsilon, \epsilon] \times B)$, for $[-M, M] \times B$ that contains the support of H_θ. The last region is $\mathbb{R}^{d+1} \setminus ([-M, M] \times B)$, over which the integrand is equal to 0.

The last region does not contribute to the integral. The contribution of the third region is bounded by $(M/\pi)^{d+1}[\eta + o(1/n)]^n$, for $\eta = \sup|\varphi_n(-v, -w)| < 1$, where the supreme is taken over the third region and it holds for all large enough n in light of (5.4). Such contribution is $o(n^{-\frac{1}{2}})$. In the second region we apply (5.5). The dominating contribution results from v and produces a bound of order $n^{-0.5\log n}$, enough to eliminate the $\sqrt{n}$ factor. For the first region we have, after multiplying by $\sqrt{n}$ and changing the variable to $(\xi, w) = (vn^{-\frac{1}{2}}, w)$:

$$\frac{\sqrt{n}}{(2\pi)^{d+1}} \int_{-\epsilon_n}^{\epsilon_n} \int_B [\varphi_n(-v, -w)]^n e^{ivx_n} \hat{H}_\theta(v, w) \, dv dw$$

$$= \frac{1}{(2\pi)^{d+1}} \int_{-\log n}^{\log n} \int_B [\varphi_n(-(\xi n^{-\frac{1}{2}}, w))]^n e^{i\xi x_n n^{-\frac{1}{2}}} \hat{H}_\theta(\xi n^{-\frac{1}{2}}, w) \, d\xi dw \ .$$

The integrand converges, for each fixed (ξ, w), to:

$$e^{-\frac{1}{2}(\xi^2 + w'\Sigma w) + i\xi x} \hat{H}_\theta(0, w) .$$

Application of the dominated convergence theorem, justified by (5.5), will give:

$$\xrightarrow[n \to \infty]{} \frac{1}{(2\pi)^{d+1}} \int_{-\infty}^{\infty} \int_B e^{-\frac{1}{2}(\xi^2 + w'\Sigma w) + i\xi x} \hat{H}_\theta(0, w) \, d\xi dw$$

$$= \phi(x)\hat{h}_\vartheta(0) \int e^{-\frac{1}{2}w'\Sigma w} \hat{k}_\eta(w) dw = \phi(x) \int h_\vartheta(y) \, dy \cdot \mathrm{E}k_\eta(Z) .$$

The last equality follows from the definition of the Fourier transform. This completes the proof of (5.3), which we will use next in order to prove the statement of the theorem.

For that proof we consider two sequences of measures on $\mathbb{R}^{d+1}$. The first is the measure that is generated by

$$\mu_n([a, b] \times A) = \sqrt{n}P(S_n - (x_n, 0)' \in [a, b] \times A) ,$$

which we wish to show that it converges to the measure generated by $\phi(x)P(Z \in A)\mu([a, b])$, for μ the Lebesgue measure. The other measure is the probability measure generated by:

$$\nu_n([a, b] \times A) = \frac{1}{\alpha_n} \int_a^b \int_A H_0(y, z)\mu_n(dy, dz) ,$$

for $\alpha_n = \sqrt{n}\mathrm{E}H_0(S_n - (x_n, 0)')$. From (5.3) it follows, for $\theta = 0$, that $\alpha_n \to \phi(x)\mathrm{E}k(Z)$ and more generally that:

$$\iint e^{i\vartheta y + i\langle \eta, z\rangle} d\nu_n(y, z) = \frac{1}{\alpha_n} \sqrt{n}\mathrm{E}H_\theta(S_n - (x_n, 0)')$$

$$\xrightarrow[n \to \infty]{} \int e^{i\vartheta y} h(y) dy \cdot \frac{\mathrm{E}[e^{i\langle \eta, Z\rangle} k(Z)]}{\mathrm{E}k(Z)} .$$

Consequently, ν_n converges in distribution to a distribution with a density that is the product of a one-dimensional Polya's distribution times a density proportional to the product of independent Polya's densities and a multivariate normal density. For the final move we apply the likelihood ratio identity, making sure that the ϵ's are selected so that the support of the numerator is contained in the part of the support of the denominator over which the denominator is strictly positive:

$$\frac{1}{\alpha_n} \mu_n([0, \delta] \times A) = \iint \frac{1_{[0,\delta]}(y) \cdot 1_A(z)}{h(y)k(z)} d\nu_n(y)$$

$$\xrightarrow[n \to \infty]{} \int \frac{1_{[0,\delta]}(y)}{h(y)} h(y) \, dy \cdot \int \frac{1_A(z)}{k(z)} k(z) f(z) \, dz = \delta \cdot P(Z \in A) ,$$

where f is the density of Z. This completes the proof of the theorem for bounded events A. Unbounded events may be intersected with a bounded event of Z probability $1 - \epsilon$. The error in the limit in assessing the probability that involves the original set is no more than $\epsilon \delta \phi(x)$. Letting $\epsilon \to 0$ deals with this general case.

The effect of having non zero values for the sequence of vectors μ_n is the addition of the term $i\langle \mu_n, w \rangle$ to the joint characteristic function, which is then reflected in the limit Gaussian density. The bound on the approximation of the joint characteristic function for all components of X_1 but the first has the addition of the $o(|\langle \mu_n, w \rangle|)$ term. These changes do not change the argument.

Finally, we would like to deal with the case of a non zero bounded function m. Choose $\epsilon > 0$ and approximate m from above by a simple function $\hat{m} = \sum_j (j\epsilon) 1_{B_j}$ such that $\hat{m} \leq m + \epsilon$. Using the partition $\{B_j\}$ we get:

$$P(S_n + m(S_n) \in (x_n, x_n + \delta) \times A) = \sum_j P(S_n + m(S_n) \in (x_n, x_n + \delta) \times A \cap B_j) .$$

Each probability in the sum can be bounded over the event B_j by:

$$P(S_n + m(S_n) \in (x_n, x_n + \delta) \times A \cap B_j)$$
$$\leq P(S_n \in (x_n - (j+1)\epsilon, x_n - j\epsilon + \delta) \times A \cap B_j) .$$

The event $A \cap B_j$ is a function of the last d coordinates of the sum. Without loss of generality, this event has a zero boundary. It follows from the proof for a zero m function that:

$$\lim_{n \to \infty} \sqrt{n} P(S_n \in (x_n - (j+1)\epsilon, x_n - j\epsilon + \delta) \times A \cap B_j)$$
$$= (\delta + \epsilon)\phi(x) P(Z \in A \cap B_j)$$

Taking a sum over j and moving to a limit establishes the upper limit:

$$\limsup_{n \to \infty} \sqrt{n} P(S_n + m(S_n) \in (x_n, x_n + \delta) \times A) \leq (\delta + \epsilon)\phi(x) P(Z \in A) .$$

A lower limit can be obtained in a similar way by replacing $\delta + \epsilon$ by $\delta - \epsilon$. The theorem follows from letting ϵ converge to 0.   □

5.5   Edge effects and higher order approximations

In this book we walk around the issue of edge effects. A justification for this cowardly avoidance is the fact that these effects are of lower order in the asymptotic derivations that we use. But lower order does not mean numerically insignificant. Frequently this is not the case and methods that take into account edge effects, even in an ad hoc manner, may produce superior numerical approximations.

The real reason for us not to include the analysis of edge effects (and other higher order effects for that matter) in this book is that we do not feel that our understanding of the issue justifies a summary in print. Although the basic principles are clear, too much work is still needed in order to produce solid statements that can be turned into computable numbers. We prefer to leave this subject open at this stage, with the hope that we, or better yet others, will be able to cover it by the time the next edition of this book is published. Yet, we may make some observations regarding the issue.

The first observation to make is that edge effects are higher order phenomena. Therefore, in order to produce a rigorous analysis of their relative contribution a higher order approximation should be applied across the board. For example, the local expansion of the global term and the evaluation in the limit of the contribution of local fluctuation should be assessed at a higher level of accuracy than is done here. At this higher level, correlations between the global term and the local fluctuations may emerge.

A situation where higher order approximations are relatively easy to obtain is the case of a smooth Gaussian field. Closely related to this case is the situation where the field is produced as a sum of independent smooth fields that converge to a Gaussian field. Indeed, an analysis in that direction was conducted for these cases in [12]. Unfortunately, these cases are already covered by the Euler characteristic method that produces an even better approximation hence the practical importance of our method in this case is questionable.

Another element that is related to higher order approximations is the consideration of parameter sets that reside in a manifold in some higher dimension domain. Again, the primary issue is to assess the effect of the curvature of the manifold on the contribution of local perturbations. As before, this is a higher order phenomenon and should be investigated in that context. Presumably, one may do so in the smooth Gaussian or almost Gaussian case. But then again, why bother? The Euler characteristic approach does it better.

Perhaps an interesting project is to consider the non smooth Gaussian case. Higher order approximations of the contribution of the global term are no different than the parallel smooth case. All that is required, and it need not be simple, is to produce a higher order evaluation of the contribution of local terms. Interesting as it may be, this goes beyond the scope of the current book.

Part II

APPLICATIONS

6

Nonparametric tests: Kolmogorov–Smirnov and Peacock

6.1 Introduction

Nonparametric inference is applied when the user is not willing to limit the model of the distribution of the data to a specific parametric family. The modeling assumption of independence between observations is kept but the marginal distribution is not specified other than making the general requirement that this marginal distribution is continuous. According to this principle, in the context of testing the null hypothesis that the observations emerge from a specific distribution, the alternative hypothesis is composed of all other continuous distributions. In the two-sample setting if the null hypothesis is that the marginal distribution in both samples is the same then this common distribution may be any distribution.

Perhaps the most popular nonparametric test for univariate observations is the Kolmogorov–Smirnov test. In the one-sample setting this test compares the empirical cumulative distribution function of the data to the theoretical cumulative distribution function. The distance between the two functions is measured in terms of the largest discrepancy and the null hypothesis is rejected if this discrepancy is above a threshold of significance. In the two-sample case the two empirical cumulative distribution functions are compared with each other in terms of the largest discrepancy. Again, the null hypothesis of equality of the two distributions is rejected if this maximal discrepancy is larger than a threshold.

Peacock's nonparametric test is an extension to higher dimensions of the Kolmogorov–Smirnov test. Just like its one-dimensional counterpart, the Peacock test compares the empirical distribution function to an hypothesized distribution

Extremes in Random Fields: A Theory and its Applications, First Edition. Benjamin Yakir.
© 2013 by Higher Education Press. All rights reserved. Published 2013 by John Wiley & Sons, Ltd.

function in the single-sample version and in the two-sample version the empirical distribution in one sample is compared with the empirical distribution in the other sample. Distances between distribution functions are measured in terms of the largest discrepancy between the two functions. The null hypothesis that states that the underlying distribution of the observations is the hypothesized one (or is the same in both samples in the two-sample version) is rejected if that largest discrepancy exceeds a critical level.

In this section we will apply our method for the approximation of the tail distribution of extremes in a random field in order to identify the critical level for rejecting the null hypothesis. We will apply it first in the context of the univariate and one-sample Kolmogorov–Smirnov test. Later we will extend the proof to include the single-sample Peacock test in two-dimensions. Indications will be given regarding the potential extension of the proof to higher dimensions still and to the two-sample setting.

The analysis of the one-dimensional Kolmogorov–Smirnov test is classical. It appears in many textbooks and is presented as an example of nonparametric statistics in any decent above-introductory-level course on statistical theory. We will give momentarily a shortened version of that representation. Knowledge of the Peacock test is less wide spread. The analysis of this test is more complex since one can no longer use simplifications that are valid in the one-dimensional case but not in higher dimensions. Still, equipped with the interpretation of statistical statements as manifestation of random fields we may identify the similarity between the one-dimensional and the multi-dimensional goodness-of-fit tests. Viewing the problem as a random field will also help us analyze natural extensions of the nonparametric test. These extensions will be discussed in the last section.

6.1.1   Classical analysis of the Kolmogorov–Smirnov test

For the one-dimensional single-sample Kolmogorov–Smirnov test we consider a sequence of i.i.d. random variables $X_1, X_2, \ldots, X_n$. The empirical cumulative distribution function that is produced by these observations is $\hat{F}(t) = \frac{1}{n} \sum_{i=1}^{n} 1_{\{X_i \leq t\}}$, for all real t. Under the assumption of continuous distribution we get that with probability 1 none of the observations share the same value. The empirical distribution function is flat between the ordered values of the observations and increases in jumps of size $1/n$ at these values. This empirical distribution is compared with the assumed underlying smooth distribution function $F(t)$. The discrepancy between the two functions is measured by the test statistic:

$$D = \sqrt{n} \max_t |\hat{F}(t) - F(t)| ,$$

and the null hypothesis is rejected for large values of D. The significance level of the test with a critical level y is given in terms of the probability $P(D \geq y)$, where the probability is computed under the assumption that the actual marginal distribution of the random variables is indeed F.

The standard analysis proceeds by showing that the distribution of D is independent of F. This is justified by the application of the transformation F to each of the observations and to the sample space itself. This transformation produces a sample $U_i = F(X_i)$ of uniformly distributed random variables. The transformation of the sample space changes the theoretical cumulative distribution function $F(t)$ to the cumulative distribution function of the uniform distribution. The reflection of this transformation on the statistic D will move the locations of the jumps and change the cumulative distribution function $F(t)$ to the cumulative distribution function of the uniform distribution. The location where D obtains its maximal value will change but the maximal value of D will not. Consequently, the analysis can proceed assuming that the original observations were taken from the uniform distribution to begin with.

In the uniform case we have that $F(t) = t$, for $0 \le t \le 1$. The random process $\{\sqrt{n}[\hat{F}(t) - t] : 0 \le t \le 1\}$ has zero expectation and a variance-covariance function that is equal to

$$n\text{Cov}(\hat{F}(t), \hat{F}(s)) = \text{Cov}\left(1_{\{U_i \le t\}}, 1_{\{U_i \le s\}}\right) = \min(s, t) - s \cdot t ,$$

for all $0 \le s, t \le 1$. This corresponds to the expectation and covariance structure of the Brownian bridge. Indeed, from the theory of weak convergence of processes we get that the limit distribution, as n goes to infinity, of the given process is the Gaussian distribution that is identified by the expectation and covariance structure that we uncovered. The conclusion is that the limit distribution of D is the same as the distribution of the maximum of the absolute value of the Brownian bridge process.

The last part in the argument concludes with the analysis of the tail of the limit distribution of D. This tail is one of the few examples in which the distribution of the maximum of a random process has an explicit known form:

$$P(D \ge y) \approx P\left(\max_{0 \le t \le 1} |B_t| \ge y\right) = 2\sum_{k=1}^{\infty}(-1)^{k+1}e^{-2k^2 y^2} \le 2e^{-2y^2} , \qquad (6.1)$$

where $\{B_t : 0 \le t \le 1\}$ is the Brownian bridge process. The approximation in the above sequence of relations reflects the approximation of the re-scaled empirical process $\sqrt{n}[\hat{F}(t) - t]$ by its Brownian bridge limit. As a matter of fact, the validity of the inequality $P(D \ge y) \le 2e^{-2y^2}$, for any value of n and $y > 0$, was proved by Massart [13], strengthening a weaker version of the statement that goes under the title of the 'Dvoretzky–Kiefer–Wolfowitz inequality'. An alternative analysis, closer in spirit to our approach, is given in [14].

In the two-sample case the two empirical distribution functions are compared with each other using the statistic:

$$\sqrt{(n_1 n_2)/(n_1 + n_2)} \max_t |\hat{F}_1(t) - \hat{F}_2(t)| ,$$

for $\hat{F}_i$ the empirical distribution function produced from a sample of size n_i, $i = 1, 2$. The convergence to a Brownian bridge still holds, providing an asymptotic justification for using the function $2 \exp\{-2y^2\}$ in order produce an approximation of the significance level of the test. However, the stronger version of the Dvoretzky–Kiefer–Wolfowitz inequality that was proved by Massart fails to hold in general [15].

6.1.2   Peacock's test

One of the goals of this chapter is to extend the analysis to the situation where the observations are multi-dimensional vectors rather than one-dimensional random variables. The statistic that we will consider was proposed by Peacock [16]. The statistic is similar to the statistic used for the Kolmogorov–Smirnov test: an empirical distribution function of the multivariate distribution of the vector is compared with a theoretical distribution in the one-sample case (or to another empirical distribution function in the two-sample case). A major difference between the two situations is that unlike the one-dimensional case, in the multi-dimensional case there is more than one natural candidate for an empirical distribution function that can be used in the comparison.

To clarify this point consider first a one-dimensional observation X_i. The evaluation of the empirical distribution that is associated with a point t on the real line corresponds to counting the number of times that the event $\{X_i \leq t\}$ occurred. Alternatively, one could have produced an empirical distribution by the consideration of events of the form $\{X_i \geq t\}$, since there is nothing special about the $-\infty$ end of the real line and distributions can be characterized by the probability above a point in the same way they are characterized by the probability below a point. However, in the one-dimensional setting we get that for either case the value of the statistic that measures the discrepancy between the resulting empirical distribution function and the theoretical probability is the same. Consequently, the Kolmogorov–Smirnov statistic does not depend on the orientation of the line.

The situation is different in the multivariate case. Specifically, consider the case where the observation $X_i = (X_{i1}, X_{i2})$ is a two-dimensional vector. A typical point in the plain is $t = (t_1, t_2)$. An empirical distribution that is associated with the point t can be constructed using events of the form $\{X_{i1} \leq t_1, X_{i2} \leq t_2\}$. However, different orientations for the two components within the vector may lead to the consideration of any one of the other three alternatives for constructing empirical distribution functions: $\{X_{i1} \geq t_1, X_{i2} \leq t_2\}$, $\{X_{i1} \leq t_1, X_{i2} \geq t_2\}$, or $\{X_{i1} \geq t_1, X_{i2} \geq t_2\}$. In this two-dimensional scenario the measure of discrepancy between the empirical distribution and its theoretical counterpart may differ, depending on the orientation that is used. Peacock's solution to this problem is to use as a test statistic the largest of the four measures of discrepancy.

The analysis of the Peacock test is more complex than the analysis required for the Kolmogorov–Smirnov test for yet another reason. For the latter test there exists a transformation of the observations and the sample space that reduces the analysis to the analysis related to the uniform distribution and, at the same time, preserves the value of the statistic. This uniform distribution has some nice features that can be exploited in the derivation of the significance level. In the multivariate case, on the other hand, no such transformation exists. Consequently, the analysis should be conducted for the original distribution and the results of the analysis depend on this distribution.

Standard tools can still be used in order to produce an approximation for the significance level of the Peacock test. Indeed, it is not too difficult to show that the empirical field that is produced by considering the discrepancy between the empirical distribution function for a given orientation and the theoretical distribution function converges, after re-scaling, to a Gaussian random field. This field is parameterized by the points of the two-dimensional Euclidean space. The heuristic analysis of the tail properties of such field is given in Example J16 of [7]. The asymptotic expansion for the Peacock test may be obtained by summing the four orientation-specific approximations, each multiplied by 2 in order to account for both tails.

However, one may find this type of analysis less than satisfactory. First, it does not incorporate explicitly the effect of the sample size since the convergence to the Gaussian limit is conducted prior to the analysis of the tail distribution. And secondly, the analysis in Aldous' book [8] is heuristic. I have no doubt that the argument is provable. Yet, it is not clear to me how to add the rigor. The method that is based on the likelihood ratio identity, we hope, does not suffer from these drawbacks.

In the next section we reanalyze the Kolmogorov–Smirnov test using our method and produce a relation between the critical level of the test and its significance value. This application is an alternative to the classical analysis that was described above. It produces somewhat weaker results – asymptotic approximations rather than exact upper bounds – however, it has the advantage of being easier to generalize to higher dimensions as required for Peacock's test. In section 6.3 we consider the Peacock's multivariate extension of the Kolmogorov–Smirnov test. The discussion there will be brief since the argument is essentially the same argument that is used for the one-dimensional case. In section 6.4 we connect the discussion to the more general problem of multivariate scanning statistics.

6.2   Analysis of the one-dimensional case

Our goal in this section is to develop a proof of (6.1) as an approximation with the aid of the method that is based on the likelihood ratio identity.

To refresh our memory we may recall that the method initiates the analysis by formulating the problem at hand in terms of a random field. The leading term in the approximation is typically associated with the largest large-deviation rate of components of the field. The collection of parameter values can be restricted by the elimination of cases where the large-deviation rate trails off. Moreover, a discrete sub-collection of the restricted set of parameters values can be selected such that the distribution of the maximum of the original random field can be approximated by the maximum of the restricted and discrete sub-collection of elements. The analysis proceeds by the identification of an alternative measure and the use of the likelihood ratio identity to transform the problem of computing a tail probability to a problem that involves the sum of expectations, each of which is associated with a particular parameter value. These expectations can be approximated by the localization theorem. Finally, the approximations are integrated in order to produce the final formula.

6.2.1   Preliminary localization

A natural candidate for the random field in the one-dimensional setting is:

$$Y_t = \sqrt{n}[\hat{F}(t) - F(t)] = n^{-\frac{1}{2}} \sum_{i=1}^{n} [1_{\{X_i \le t\}} - P(X_i \le t)],$$

defined for $-\infty < t < \infty$. The expectation of a component Y_t is equal to zero and the variance is equal to $F(t)[1 - F(t)]$. A normal approximation of the binomial distribution may trigger one to suspect that

$$\log P(Y_t \ge y) \approx \log \bar{\Phi}(y/\{F(t)[1 - F(t)]\}^{\frac{1}{2}}) \approx -\frac{y^2}{2F(t)[1 - F(t)]}.$$

As a function of t, this first-order approximation of the marginal tail distribution of Y_t is maximized at the median $\hat{t} \iff F(\hat{t}) = 0.5$, where the approximation of the marginal obtains the value e^{-2y^2}.

A refined analysis can be obtained by the examination of marginal probabilities in the vicinity of the median $\hat{t}$. The reduction in the exponent between the probability of the tail at t and the probability at $\hat{t}$ is:

$$\frac{y^2}{2}\left\{\frac{1}{1/4} - \frac{1}{F(t)[1 - F(t)]}\right\} = \frac{-2y^2(2F(t) - 1)^2}{1 - (2F(t) - 1)^2} \approx -8[f(\hat{t})]^2[y(t - \hat{t})]^2.$$

Consequently, one may expect to be able to restrict maximization of the test statistic to the range $T = \hat{t} \pm C/y$, for some large but fixed value C, without a great loss in the resulting one-sided probability of exceeding the critical level.

A more careful analysis will inspect the large deviation rate for the centered and re-scaled binomial random variable Y_t as a function of both n and t instead of relying on the limit normal approximation. Here we will satisfy ourselves with the current heuristic argument until we introduce the appropriate log-moment generating function which can be used in order to identify the rates. In the final integration step we have a representation that includes a large deviation factor that is produced by these log-moment generating functions. These factors are the terms that should have been used in the argument that justified the preliminary localization step.

6.2.2   An approximation by a discrete grid

The second ingredient in the preliminary preparation for the application of the likelihood ratio identity is the approximation, based on a dense but finite sub-collection of values of the parameter, of the maximal value of the field. An attempt to mimic the analysis that was used in Section 3.2.2 for the scanning statistic will call for the analysis of an upper bound of the form:

$$e^{2y^2} \sum_{t \in \hat{T}} P(Y_t \le y - \frac{\epsilon}{y}, \max_{t \le s \le t + \delta/y^2} Y_s \ge y) , \qquad (6.2)$$

where $\hat{T} \subset T$ is the dense sub-collection that is used in the approximation of the maximization over T.

An assessment of the bound in (6.2) can be carried out by the conditioning on the value of Y_t and an application of an inequality in the spirit of Fernique's inequality that is stated in Theorem 3.1. The proof of Fernique's inequality relies on a chaining argument and bounds on the tails of marginal distributions of increments. Originally, the argument was applied to a Gaussian tail but the basic idea may be used for other types of tail behavior. A generalized argument is given in theorem [17] with their main result quoted in Theorem A.5 in the appendix. Unfortunately, we cannot use the theorem as stated since, as Chernoff's bound (Theorem A.4) demonstrates, the tail of the binomial distribution is sub-Gaussian with one order for the initial part of the tail and with a different order for the final part of the tail. Still, one can apply the chaining argument and obtain a useful bound on the tail of extreme values.

The minimal value of the random variable Y_t is $u_0 = -\sqrt{n} F(t)$. Denote $T_t = [t, t + \delta/y^2]$. Conditioning on the event $\{Y_t = u\}$, for u between u_0 and $y - \frac{\epsilon}{y}$, gives:

$$P\left(Y_t \le y - \frac{\epsilon}{y}, \max_{s \in T_t} Y_s \ge y\right) = \sum_{u=u_0}^{y-\frac{\epsilon}{y}} P(Y_t = u) P(\max_{s \in T_t} Y_s - Y_t \ge y - u | Y_t = u) .$$

This summation can be approximated by a Gaussian integration after the production of a bound on the conditional tail probability, since the probability mass function of Y_t can be approximated by a normal density. Some care should be given to small values of u, but for such values the conditional tail probability will be extremely small.

The maximum of the incremental process over the interval T_t can be bounded using a chaining argument. A basic ingredient for the application of this argument is a bound on the tail of increments:

$$Y_s - Y_r = \{N_{s,r} - n F_{s,r})\}/\sqrt{n} ,$$

for $N_{s,r}$ the count and $F_{s,r}$ the probability associated with the interval $(s, r]$, $s, r \in T_t$. It should be noted that the conditional distribution of $N_{s,r}$, given that $Y_t = u$, is binomial that counts the number of successes among $n_u = n(1 - F(t)) - u\sqrt{n} \approx n/2$ independent trials with a probability

$$F_{s,r|t} = [F(r) - F(s)]/[1 - F(t)] \approx 2f(t)(r - s)$$

of success. An increment can be represented in the form:

$$Y_s - Y_r = \{N_{s,r} - n_u F_{s,r|t}\}/\sqrt{n} - u F_{s,r|t} \,.$$

Observe that the conditional variance of an incremental count is:

$$\mathrm{Var}(N_{s,r}|Y_t = u) = n_u F_{s,r|t}(1 - F_{s,r|t}) \approx nf(t)(r - s) \,.$$

Consequently, if we apply the inequality (Theorem A.4) to an increment we get:

$$P(Y_r - Y_s \geq x|Y_t = u) = P(N_{s,r} - n_u F_{s,r|t} \geq \sqrt{n}(x + u F_{s,r|t})|Y_t = u)$$

$$\lesssim \exp\left\{ -\frac{(x + 2uf(t)(r - s))^2}{2f(t)(r - s)} \frac{1}{1 + \frac{2u}{3\sqrt{n}} + \frac{x}{3\sqrt{n}f(t)(r-s)}} \right\} \,.$$

At the heart of the chaining argument is the representation of a typical increment $Y_s - Y_t$, for $s \in T_t$, as a telescopic sum:

$$Y_s - Y_t = \sum_{l=1}^{\infty}(Y_{t_l(s)} - Y_{t_{l-1}(s)}) \,,$$

with $t_0(s) = t$ and $t_l(s)$ the central point of the interval of length $2^{-l}\delta/y^2$ in a partition of T_t that has 2^l elements. The interval that has $t_l(s)$ as a center covers the point s. Notice that $t_l(s) - t_{l-1}(s) = 2^{-(l+1)}\delta/y^2$ and denote $t_{j,l} = t + j\delta 2^{-l+1}/y^2$. Let $\pi_l = (1 - q)q^{l-1}$, for $1/\sqrt{2} < q < 1$. Observe that $y - u = \sum_{l=1}^{\infty}(y - u)\pi_l$.

An illustration of the chaining argument is given in Figure 6.1. It describes a pedigree that is constructed for the sake of producing a bound. The founder of the pedigree is located at point t. Each diagonal segment is associated with an increment between a parent and an offspring. The horizontal distance of the diagonal segment is the distance between the parameter value for the parent and the parameter value for the offspring. The vertical distance is the fraction of the total distance that is allocated for the increment between the parent and the offspring. In order for an increment between a point in the parameter space and the founder to exceed x at least one of the parent–offspring pairs must exceed

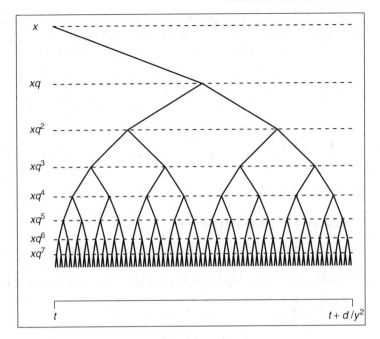

Parameter

Figure 6.1   An illustration of the chaining argument. Diagonal segments correspond to a parent–offspring pair in a chain that connects any point with the founder of the pedigree.

the allocated distance. Hence, the probability that there is a point where the increment to the the founder is more than x is contained in the countably infinite union of events that a parent–offspring pair exceeds the allocated distance.

It follows, from the application of the upper bound on the tail of an increment, that:

$$P\left(\max_{s \in T_t} Y_s - Y_t \geq y - u \mid Y_t = u\right)$$

$$\leq \sum_{l=1}^{\infty} \sum_{j=1}^{2^{l+1}} P\left(Y_{t_{j,l}} - Y_{t_{j-1,l}} \geq (y-u)\pi_l \mid Y_t = u\right)$$

$$\approx \sum_{l=1}^{\infty} 2^l \exp\left\{-\frac{(y(y-u)(q\sqrt{2})^l(1/q - 1) + (u/y)f(t)2^{-\frac{l}{2}}\delta)^2}{\delta f(t)\left[1 + \frac{2u}{3\sqrt{n}} + \frac{y^2(y-u)(2q)^l(1/q-1)}{1.5\sqrt{n}f(t)\delta}\right]}\right\}.$$

We complete the analysis by multiplying the bounds by $\exp\{2y^2\}$ times the density of Y_t and summation over $\hat{T}$. The summation is approximated by integration and a change of variable $x = y(y - u)$ is used in order to produce:

$$(6.2) \lesssim \sum_{l=1}^{\infty} |\hat{T}|2^l \int_{u_0}^{y-\frac{\epsilon}{y}} \exp\left\{4y(y-u) - \frac{(y(y-u)(q\sqrt{2})^l(1/q-1) + \frac{u}{y}f(t)2^{-\frac{l}{2}}\delta)^2}{\delta f(t)\left[1 + \frac{2u}{3\sqrt{n}} + \frac{y^2(y-u)(2q)^l(1/q-1)}{1.5\sqrt{n}f(t)\delta}\right]}\right\} du$$

$$= \sum_{l=1}^{\infty} \frac{2C2^l}{\delta} \int_{\epsilon}^{y(y-u_0)} \exp\left\{4x - \frac{(x(q\sqrt{2})^l(1/q-1) + (1-x/y^2)f(t)2^{-\frac{l}{2}}\delta)^2}{\delta f(t)\left[1 + \frac{2(1-x/y^2)}{3\sqrt{n}/y} + \frac{x(2q)^l(1/q-1)}{1.5\sqrt{n}/yf(t)\delta}\right]}\right\} dx ,$$

which converges to 0 when $\delta \to 0$.

6.2.3 Measure transformation

The alternative distributions that are used for the production of the likelihood ratio identity rely on the fact that Y_t is a linear transformation of binomial random variables. Recall that the log-moment generating function for the Bernoulli variable is $\varphi_p(\xi) = \log(pe^\xi + 1 - p)$, where p is the probability of success. It follows that the log-moment generating function for Y_t is

$$\psi_t(\xi) = n\left[\varphi_p\left(\xi/n^{\frac{1}{2}}\right) - \left(\xi/n^{\frac{1}{2}}\right)\varphi_p'(0)\right] ,$$

where $\varphi_p'(0) = p = p_t = F(t)$.

The large deviation rate is associated with the value of ξ that solves the equation

$$\psi_t'(\xi) = n^{\frac{1}{2}}\left[\varphi_p'\left(\xi/n^{\frac{1}{2}}\right) - \varphi_p'(0)\right] = y$$

$$\Rightarrow \quad \xi = \xi_t \approx y/\varphi_p''(0) = y/[F(t)(1 - F(t))] .$$

The resulting approximation of the log of the marginal probability, the approximation that agrees to the first order with the normal approximation that we used above in order to restrict the parameter set to an interval about the median, takes the form:

$$\psi_t(\xi_t) - \xi_t\psi_t'(\xi_t) = n\left[\varphi_p\left(\xi_t/n^{\frac{1}{2}}\right) - \left(\xi_t/n^{\frac{1}{2}}\right)\varphi_p'\left(\xi_t/n^{\frac{1}{2}}\right)\right] .$$

The value of t that maximizes this quantity is denoted $\hat{t}$. Note that for $t = \hat{t}$ we obtain that $\xi_{\hat{t}} \approx 4y$.

Let T now be the grid with span δ/y^2 of points in the interval $\hat{t} \pm C/y$. Associate with each $t \in T$ a log-likelihood ratio of the form $\ell_t = \xi_t Y_t - \psi_t(\xi_t)$. With the aid of such log-likelihood ratios one may produce the likelihood ratio identity:

$$P\left(\max_{t\in T} Y_t \geq y\right) = \sum_{t\in T} E_t\left(\frac{1}{\sum_{s\in T} e^{\ell_s}} ; \max_{s\in T} Y_s \geq y\right) ,$$

where P_t is the alternative distribution that is associated with the log-likelihood ratio ℓ_t.

A local random field is produced by the consideration of differences between the log-likelihood ratio at t and the log-likelihood ratios at other parameter values. The components of the local field are:

$$\ell_s - \ell_t = \xi_s Y_s - \xi_t Y_t - (\psi_s(\xi_s) - \psi_t(\xi_t)) .$$

This local field serves to produce the summation and maximization statistics:

$$S_t = \sum_{s \in T} e^{\ell_s - \ell_t} , \quad \text{and} \quad M_t = \max_{s \in T} e^{\ell_s - \ell_t} .$$

Combine that with the re-centered log-likelihood ratio:

$$\tilde{\ell}_t = \xi_t(Y_t - \psi_t'(\xi_t)) ,$$

in order to get a representation of the summands in the log-likelihood identity that is convenient for the application of Theorem 5.2:

$$\mathrm{E}_t\left(\frac{1}{\sum_{s \in T} e^{\ell_t}}; \max_{s \in T} Y_s \geq y\right) \approx e^{\psi_t(\xi_t) - \xi_t \psi_t'(\xi_t)} \mathrm{E}_t\left(\frac{M_t}{S_t} e^{-[\tilde{\ell}_t + m_t]}; \tilde{\ell}_t + m_t \geq 0\right) ,$$

where $m_t = \log M_t$.

6.2.4 The asymptotic distribution of the local field and the global term

The null distribution P assigns the marginal density f to the independent observations $X_1, \ldots, X_n$. The distribution P_t corresponds to an alternative distribution over the same collection of observations. The observations under the alternative distribution are again i.i.d., but the marginal density is tilted and takes the form
$$f_t(x) = f(x) \exp\{\xi_t(1_{\{x \leq t\}} - F(t))/n^{\frac{1}{2}} - \psi_t(\xi_t)/n\} .$$
In order to apply the localization theorem we need to identify the local limit distribution of $\tilde{\ell}_t$ and of the local field $\{\ell_s - \ell_t : s \in T\}$ and prove asymptotic independence between them. The analysis of the limit distributions should be carried out in the context of the alternative distribution P_t. This analysis is simplified substantially by the fact that both the centered log-likelihood ratio $\tilde{\ell}_t$ and the local field are produced as linear functions of the collection of independent random fields:

$$\{1_{\{X_i \leq s\}} : s \in T\} , \quad i = 1, 2, \ldots, n .$$

Due to this fact we are in a position to apply Theorem 5.3 and get that the limit joint distribution of the local field and the global term is Gaussian. Computation of the expectation and covariance structure are sufficient for the specification of the limit local field and the proof of asymptotic independence.

For the analysis of the local field one may examine the representation:

$$Y_s = \sum_{i=1}^{n} \left[1_{\{X_i \le s\}} - F(s) \right] n^{-\frac{1}{2}} .$$

Let $y = y_n$ diverge to infinity with the increase in the sample size. The expectation of the sum Y_s is the sum of the expectations. Consequently, the expectation of a component in the sum under the alternative distribution is:

$$n^{-\frac{1}{2}} E_t \left[1_{\{X_i \le s\}} - F(s) \right]$$

$$= n^{-\frac{1}{2}} E \left[\left(1_{\{X_i \le s\}} - F(s) \right) e^{\frac{\xi_s}{\sqrt{n}} \left(1_{\{x \le t\}} - F(t) \right) - \psi_t(\xi_s)/n} \right]$$

$$\approx \frac{4y}{n} E \left[\left(1_{\{X_i \le s\}} - F(s) \right) \left(1_{\{X_i \le t\}} - F(t) \right) \right]$$

$$= \frac{4y}{n} \mathrm{Cov} \left(1_{\{X_1 \le s\}}, 1_{\{X_1 \le t\}} \right) ,$$

where the approximation is valid if $n \to \infty$ and $y/n^{\frac{1}{2}} \to 0$. In the approximation we used the fact that $e^x \approx 1 + x$, for small x, and the fact that $\xi_s \approx 4y$. From this approximation we may conclude that for large values of n:

$$E_t \left[\xi_s Y_s - \xi_t Y_t \right] \approx 16 y^2 \left[\mathrm{Cov} \left(1_{\{X_1 \le s\}}, 1_{\{X_1 \le t\}} \right) - \mathrm{Var} \left(1_{\{X_1 \le t\}} \right) \right] .$$

The deterministic part of the local field is $-\psi_s(\xi_s) + \psi_t(\xi_t)$. However, for large values of n and y:

$$\psi_s(\xi_s) = n \left[\varphi_p \left(\xi_s/n^{\frac{1}{2}} \right) - \left(\xi_s/n^{\frac{1}{2}} \right) \varphi_p'(0) \right] \approx 16 y^2 \frac{1}{2} \mathrm{Var} \left(1_{\{X_1 \le s\}} \right) ,$$

with a similar approximation holding for $\psi_t(\xi_t)$. Therefore,

$$\psi_s(\xi_s) - \psi_t(\xi_t) \approx 8 y^2 \left[\mathrm{Var} \left(1_{\{X_1 \le s\}} \right) - \mathrm{Var} \left(1_{\{X_1 \le t\}} \right) \right] .$$

Combining the approximation for the expectation of the random, the approximation of the deterministic part, and expression for the variance and covariance of Bernoulli random variables we obtain:

$$E_t(\ell_s - \ell_t) \approx 16 y^2 [F(s \wedge t) - F(s) F(t)$$

$$- \frac{1}{2} \{ F(s)(1 - F(s)) + F(t)(1 - F(t)) \}]$$

$$= -\frac{1}{2} 16 y^2 [F(s \vee t) - F(s \wedge t) - (F(s) - F(t))^2] .$$

Specifically, for $s = t + j\delta/y^2$, with j that is at most slowly changing as a function of y, we get:

$$E_t(\ell_s - \ell_t) \approx -8f'(t)\delta|j| \approx -8f'(\hat{t})\delta|j|. \tag{6.3}$$

The variance of $\ell_s - \ell_t$ involves a sum of variances. Straightforward computations, and the observation that the variances of components of the random field under the alternative distribution are asymptotic to the variance under the null distribution, will lead to the approximation:

$$\text{Var}_t(\ell_s - \ell_t) \approx 16y^2[F(s \vee t) - F(s \wedge t) - (F(s \vee t) - F(s \wedge t))^2].$$

As a result we get that:

$$E_t(\ell_s - \ell_t) \approx -\frac{1}{2}\text{Var}_t(\ell_s - \ell_t). \tag{6.4}$$

The analysis of the covariance between $\ell_s - \ell_t$ and $\ell_r - \ell_t$ is similar and yields in the case $t < r < s$:

$$\text{Cov}_t(\ell_s - \ell_t, \ell_r - \ell_t) \approx 16y^2[F(r) - F(t) - (F(s) - F(t))(F(r) - F(t))].$$

Letting $y \to \infty$ will unfold the fact that the asymptotic distribution of $\{\ell_s - \ell_t\}$, for $s = t + j\delta/y^2$ and $|j|$ not too large, is a two-sided Gaussian random walk with a negative drift. The variance of an increment of that random walk is $16f'(\hat{t})\delta$.

Under the alternative distribution we get that the expectation of $\tilde{\ell}_t$ is 0 and the variance is $16y^2\psi_t''(\xi_t) = 16y^2\varphi_p''(\xi_t/n^{\frac{1}{2}})$, for $p = F(t)$. If we take $\kappa = y^2$ for the purpose of Theorem 5.2 and if $y/n^{\frac{1}{2}} \to 0$ then we get that $\mu = 0$ and, since $t \to \hat{t}$ and thus $\varphi_p''(\xi_t/n^{\frac{1}{2}}) \to \varphi_{0.5}''(0) = 1/4$, that $\sigma^2 = 16/4 = 4$.

The validation of the asymptotic independence that is needed for Theorem 5.3 follows from the fact the the covariance between $\ell_s - \ell_t$ and $\tilde{\ell}_t$ is of the order of a constant. However, the standard deviation of $\tilde{\ell}_t$ diverges to infinity proportionally to y. Consequently, the correlation between the global term and elements of the local field tends to zero.

6.2.5   Application of the localization theorem and integration

The two-sided Gaussian random walk with a negative drift (that equals half the variance) has emerged in previous examples. In particular, we encountered it in the case of a scanning statistic and in the context of the sequential test for a normal mean. In both cases, the limit of the expectation of the ratio between the maximum and the sum, the constant that summarizes the effect of local fluctuations on the probability that the random field will obtain a large value, is associated with the overshoot function ν that is approximated in (2.4):

$$E[\mathcal{M}/\mathcal{S}] = 8f'(\hat{t})\delta \cdot \nu\left([16f'(\hat{t})\delta]^{\frac{1}{2}}\right). \tag{6.5}$$

The conclusion of Theorem 5.2 in the current setting is that

$$\mathrm{E}_t\left(\frac{M_t}{S_t}e^{-[\tilde{\ell}_t+m_t]}; \tilde{\ell}_t + m_t \geq 0\right) \approx \delta\frac{8f'(\hat{t})}{y\sqrt{8\pi}} \cdot v\left([16f'(\hat{t})\delta]^{\frac{1}{2}}\right),$$

which leads to the approximation:

$$P\left(\max_{t\in T} Y_t \geq y\right) \approx \delta\frac{8f'(\hat{t})}{y\sqrt{8\pi}} \cdot v\left([16f'(\hat{t})\delta]^{\frac{1}{2}}\right)\sum_{t\in T} e^{\psi_t(\xi_t)-\xi_t\psi_t'(\xi_t)}$$

In order to obtain a formula for the situation where normal approximation is valid for the large variation components we may identify the limit of each exponent in the sum and then expand it as a function of t in the vicinity of $\hat{t}$:

$$\psi_t(\xi_t) - \xi_t\psi_t'(\xi_t) \approx -\frac{\xi_t^2}{2}\psi_t''(0) = -\frac{y^2}{2\psi_t''(0)},$$

since $\xi_t \approx y/\psi_t''(0)$. We use the fact that $\psi_t''(0) = F(t)[1 - F(t)]$, the fact that

$$F(t)[1 - F(t)] = (0.5 + [F(t) - 0.5])(0.5 - [F(t) - 0.5])$$

$$= 0.25 - [F(t) - 0.5]^2,$$

and the fact that $F(t) - 0.5 = F(t) - F(\hat{t}) \approx f'(\hat{t})(t - \hat{t})$ to obtain:

$$\psi_t(\xi_t) - \xi_t\psi_t'(\xi_t) \approx -2y^2 - 8[f(\hat{t})]^2[y(t - \hat{t})]^2.$$

Upon plugging this approximation in the expression for the probability of and extreme value and setting $t = \hat{t} + j\delta/y^2$ we get:

$$P\left(\max_{t\in T} Y_t \geq y\right) \approx e^{-2y^2} \cdot v\left([16f'(\hat{t})\delta]^{\frac{1}{2}}\right)\frac{2f'(\hat{t})\delta/y}{\sqrt{2\pi}}\sum_{j:|j|\leq yC/\delta} e^{-\frac{1}{2}[2f(\hat{t})j\delta/y]^2}.$$

The summation is approximated, for large values of y, by an integral of a Gaussian kernel which gives $2\Phi(C/[2f(\hat{t})]) - 1$, for Φ the standard normal cumulative distribution function. Letting $C \to \infty$ and $\delta \to 0$, and taking into account the fact that the limit of the function v at the origin is equal to 1, one gets the approximation:

$$P\left(\max_t Y_t \geq y\right) \approx e^{-2y^2}.$$

For large values of y the probability that the maximum of the absolute value exceeds y is asymptotic to twice the one-sided probability. Consequently, the resulting approximation is in agreement with the upper bound (6.1).

In the two-sample Kolmogorov–Smirnov test one examines the empirical distribution functions $\hat{F}_j(t) = (1/n_j)\sum_{i=1}^{n_j} 1_{\{X_{ij}\leq t\}}$, for $j = 1, 2$, the two samples. The test statistic maximizes the absolute value of:

$$Y_t = \sqrt{\frac{n_1 n_2}{n_1 + n_2}}(\hat{F}_1(t) - \hat{F}_2(t)).$$

The maximal marginal probability again takes place at $\hat{t}$ the median of the common distribution function F. It is convenient to assume that the sample sizes are proportional to one another and both converge to infinity. Alternative measures are produced by tilting the statistics Y_t. Examination of local behavior of increments can be carried out via the representation:

$$Y_r - Y_s = \sqrt{\frac{n_1 n_2}{n_1 + n_2}}\left(\hat{F}_1(r) - \hat{F}_1(s) - F_{s,r}\right) - \sqrt{\frac{n_1 n_2}{n_1 + n_2}}\left(\hat{F}_2(r) - \hat{F}_2(s) - F_{s,r}\right),$$

and the bounding or approximation of each of the two terms. The entire derivation is a nice exercise for the reader to carry out.

6.2.6   Checking the conditions of the localization theorem

For the current example we can use the simplified version of Theorem 5.1. Still, just like we did for the two examples that were used for demonstration in Chapter 3, we still prefer to prove the conditions of the more complex Theorem 5.2.

In the context of the localization theorem we want to define a local σ-algebra and identify the local approximations of the terms that measure the effect of local fluctuations. It is convenient to use for that purpose the σ-algebra that is generated by the finite collection of increments: $\{\ell_s - \ell_t : |s - t| \leq \tau/y^2\}$. Then $\hat{M}_t$ and $\hat{S}_t$ are the maximization and the summation, respectively, of likelihoods that are restricted to this finite collection.

Condition I* holds trivially with $C = 1$. For Condition V* we have that

$$P_t\left(M_t > e^{g(y^2)}\right) = P_t\left(\max_{\{s:|s-\hat{t}|\leq C/y\}} e^{\ell_s - \ell_t} > e^{g(y^2)}\right) \leq (2Cy/\delta)e^{-g(y^2)},$$

so taking $g(y^2) = (1 + \epsilon)\log y$ will do the job. Conditions III* and IV* follow from the application of the local limit Theorem 5.3 so we are left only with the need to check Condition II*.

Using the same argument that was used in the proof of Condition II* for the examples that were analyzed in Chapter 3, and in particular the example of an indicator function serving as a kernel, we can see that it is sufficient to investigate the probabilities:

$$P_t(\{\ell_s - \ell_t \geq \log(\epsilon\, p_s)\} \cap \{\tilde{\ell}_t + \hat{m}_t \in x + (0, \delta]\})\,,$$

for $|s - t| > \tau/y^2$. As in the case of the scanning statistic, we split the analysis between the case where $\tau/y^2 < |s - t| < (\log y)^2/y^2$ and the case where $(\log y)^2/y^2 < |s - t|$ and use uniform weights for the former region and exponentially decreasing weights for the latter region. More tedious details go into the proof but no new ideas.

6.3   Peacock's test

Peacock's test generalizes the Kolmogorov–Smirnov test to higher dimensions. The setting where this test can be applied involves a sample $X_1, \ldots, X_n$ of independent random points on a d-dimensional plane. Henceforth we will assume that $d = 2$. However, the generalization to a finite dimension that is larger than 2 is straightforward.

Denote the distribution of X_i on the plane by F and define the empirical distribution over Borell subsets of T:

$$\hat{F}(A) = \frac{1}{n}\sum_{i=1}^{n}1_{\{X_i \in A\}}.$$

The Peacock nonparametric goodness-of-fit test is based on the difference between the empirical and theoretical distribution functions, applied to different quadrants. Specifically, let $t = (t_1, t_2)$ be a point in the support of F and consider the four closed quadrants with respect to this point by:

$$Q_1(t) = (\infty, t_1] \times (\infty, t_2]\,, \quad Q_2(t) = (\infty, t_1] \times (-\infty, t_2]\,,$$
$$Q_3(t) = (-\infty, t_1] \times (-\infty, t_2]\,, \quad Q_4(t) = (-\infty, t_1] \times (\infty, t_2].$$

The re-scaled difference between the empirical distribution and the probability assigned to the set $Q_j(t)$ is $Y_{j,t} = \sqrt{n}\{\hat{F}(Q_j(t)) - F(Q_j(t))\}$. The test statistic is based on the largest discrepancy between these empirical distribution functions:

$$\max_{j\in\{1,2,3,4\}}\ \max_{t\in T}|Y_{j,t}|.$$

The null hypothesis of validity of the assumed distribution F is rejected if this statistic is larger than a critical level. If the critical level is equal to y we get that the significance level of the test is:

$$P\left(\max_{j\in\{1,2,3,4\}}\ \max_{t\in T}|Y_{j,t}| \geq y\right) \approx 2\sum_{j=1}^{4}P\left(\max_{t\in T}Y_{j,t} \geq y\right),$$

NONPARAMETRIC TESTS   121

where the approximation is appropriate for large values of y. Our goal is to produce an approximation for the probabilities that appear in the sum.

We will derive an approximation for the case $j = 1$. Hence, in the following we eliminate the index j and denote $\hat{F}(t) = \hat{F}(Q_1(t))$ and $F(t) = F(Q_1(t))$.

The random field in the current example is two-dimensional. The marginal distributions that are associated with elements Y_t of the field are binomial, the same as they were in the one-dimensional setting of the Kolmogorov–Smirnov test. Their large deviation rates can assessed via the examination of the terms

$$\psi_t(\xi_t) - \xi_t \psi_t'(\xi_t) = n \left[\varphi_p \left(\xi_t / n^{\frac{1}{2}} \right) - \left(\xi_t / n^{\frac{1}{2}} \right) \varphi_p' \left(\xi_t / n^{\frac{1}{2}} \right) \right],$$

where $\varphi_p(\xi)$ is the log-moment generating function of the Bernoulli random variable with $p = F(t)$ the probability of success and $\xi_t \approx y/[p(1 - p)]$ is the value ξ that solves the equation:

$$\psi_t'(\xi) = n^{\frac{1}{2}} \left[\varphi_p' \left(\xi / n^{\frac{1}{2}} \right) - \varphi_p'(0) \right] = y.$$

Again, we would like to consider maximization of this rate. This maximization occurs when $p = 1/2$ and leads to the same rate that is introduced in the one-dimensional case. However, unlike the one-dimensional case, the collection of t values that obtains this maximization is not a single point but, instead, it is a curve in the plane:

$$\hat{t} = \{t : F(t) = 1/2\}.$$

It should be noted that this curve is compact. Due to the implicit function theorem the curve is also smooth if the density of the underlying distribution is smooth.

The preliminary localization step narrows the range of interesting parameter values to a tube about this curve:

$$\{t : |F(t) - 1/2| \le C/y\},$$

for a large enough C.

The approximation of the maximal statistic by the maximizer of a discrete two-dimensional field is obtained via the bounding of probabilities such as:

$$P\left(Y_t \le y - \frac{\epsilon}{y}, \max_{s \in T_t} Y_s \ge y \right),$$

where currently $T_t = [t_1, t_1 + \delta/y^2] \times [t_2, t_2 + \delta/y^2]$. These probabilities are the translation to the two-dimensional setting of the probabilities that are given in the sum (6.2) that was analyzed in the one-dimensional case. Bounding this probability can be carried out by conditioning on the value of Y_t and producing a Fernique's type of inequality for the tail of the conditional incremental random field. The proof of the inequality itself follows the proof that was developed for the one-dimensional case word for word. The only difference here is that each parent point has four offspring instead of the two that are used in the one-dimensional scenario. The four offspring are located at the centers of the four quadrants that divide the square that has the parent as its center.

The fact that a partition to four quadrants is carried out changes the number of elements in the lth level of the partition from 2^l to 4^l. Another difference is the change in the definition of the probability $F_{r,s} = F(r) - F(s)$ from a probability of an interval to a probability of a slightly more complex set: $Q(r)\backslash Q(s)$. The approximation of this probability for r in the vicinity of s, is $F_{r,s} \approx \dot{F}(s)'(r-s)$, for $\dot{F}(s)$ the gradient of the cumulative distribution function, evaluated at the point s.

The alternative distribution and the measure transformation are unaffected by the change in dimensionality. The localization theorem can be quoted essentially verbatim from the proof that is developed for the one-dimensional case. A slight modification appears in the step that identifies the limit of the local field and the resulting evaluation of the term that measures the effect of local fluctuations. In the current setting one obtains for $s = (s_1, s_2) = (t_1 + j_1\delta/y^2, t_2 + j_2\delta/y^2)$ that the limit variance of the corresponding element of the local field is:

$$\mathrm{Var}(\ell_s - \ell_t) \approx 16\dot{F}_1(t)\delta|j_1| + 16\dot{F}_2(t)\delta|j_2| \, ,$$

for $\dot{F}_1$ and $\dot{F}_2$ the first and second components of the gradient of F. The limit covariance structure of the local field suggests that limit of the local field corresponds to a sum of two independent two-sided random walks. These random walks have negative drifts that equal half the variances. The conclusion regarding the term that measures the contribution of local fluctuations is that:

$$\mathrm{E}[\mathcal{M}/\mathcal{S}] = \delta^2 \prod_{i=1}^{2} \left\{ 8\dot{F}_i(t) \cdot v \left([16\dot{F}_i(t)\delta]^{\frac{1}{2}} \right) \right\}. \tag{6.6}$$

Consequently, since the distribution of $\tilde{\ell}_t$ is unaffected by the change in dimensionality, the approximation that results from Theorem 5.2 is:

$$\mathrm{E}_t \left(\frac{M_t}{S_t} e^{-[\tilde{\ell}_t + m_t]}; \tilde{\ell}_t + m_t \geq 0 \right) \approx \mathrm{E}[\mathcal{M}/\mathcal{S}]/(y\sqrt{8\pi}).$$

Finally, the conclusion from the integration step is that:

$$P\left(\max_{t \in T} Y_t \geq y \right) \approx \prod_{i=1}^{2} \sum_{t \in T} \left\{ 8\dot{F}_i(t) \cdot v \left([16\dot{F}_i(t)\delta]^{\frac{1}{2}} \right) \right\} \cdot \frac{\delta^2}{y\sqrt{8\pi}} e^{\psi_t(\xi_t) - \xi_t \psi_t'(\xi_t)}.$$

With the increase of the threshold to infinity we get that the sum converges to an integral. An approximation of the original probability may be obtained by sending C to infinity and δ to zero.

6.4 Relations to scanning statistics

The Peacock test is an example of analysis of a scanning statistic on a two-dimensional space. The type of signals that one attempts to discover are quadrants of the plane and the source of information and randomness is the empirical distribution. The same principles that were applied for the analysis of this test are relevant to other types of scanning statistics. The results of the analysis tend to share similarities.

An immediate extension of the Peacock test will consider empirical distributions of rectangular regions of the form $[t_1, s_1] \times [t_2, s_2]$. Such regions are parameterized by pairs of points (t, s). Hence, the parameter space is a subset of $\mathbb{R}^4$. Maximizing the re-scaled absolute difference between the empirical distribution of rectangles and their theoretical probabilities, where the maximization is taken over the parameter space, forms the basis for an alternative goodness-of-fit test. This type of test is described as Example J17 in [7].

The analysis of this extension should be a simple exercise for the reader who completed the details for the different variants of the Kolmogorov–Smirnov and the Peacock tests that were discussed in the previous sections. Again, the exponent of the large deviation rate is associated with rectangles of probability 1/2 and is asymptotic to $-2y^2$, where y is the threshold. For the term that measures the local fluctuations we need to change the power in (6.6) from 2 to 4. Other minor details need to be modified, but the modifications are trivial.

This last example has a parallel formulation in the context of scanning statistics that generalize the basic example of Section 2.3 to situations where the signal emerges in a two-dimensional, or more generally d-dimensional, space. One difference between this new generalization and the goodness-of-fit test is the source of randomness. In the scanning statistic setting the source can be modeled as a higher dimensional white noise, Gaussian or not, instead of an empirical distribution. The choice of the source will affect the final statement of the outcome, but will have less of an effect on the method for obtaining that statement. Again, the large deviation rate will be associated with values of the parameter that maximize the marginal variance assuming that the expectation was taken care of. If the signal is in the form of an indicator of a rectangle then the term that is associated with local fluctuations will take the form of an appropriate product of terms associated with one-dimensional scanning. If the signal is of the form of a smooth kernel then the field is smooth and the resulting term is associated with the determinant of the Hessian that is associated with derivatives of the covariance function.

Another difference will be shown if the scanning statistic is standardized before the maximization over the parameter space. This situation corresponds more directly with the scanning statistics of Section 2.3. In this case the variance is constant across the parameter space and equals 1. The resulting exponent of large deviation will be fixed in the Gaussian case (and equal to $-y^2/2$) and will be almost fixed in non-Gaussian cases that are asymptotically Gaussian. The constant that summarizes the local fluctuations will be different, as well as the polynomial terms. Explicit formulae are known for the smooth case and for rectangular kernels with different characteristics. The details for some examples can be found in [18]. For kernels with nonrectangular shapes that produce a nonsmooth local field, say circles, one may write formally the term that is associated with local fluctuation as the expected ratio of a maximal exponentiated field over the sum of the same exponentiated field. However, we do not know of an explicit formula for this term and leave it as a challenge for the ambitious reader.

An even more challenging problem is to analyze the scanning statistics that are associated with shapes that are less regular than rectangles or circles. With such shapes the dimension of the parameter space may grow, a situation which is challenging, but not hopeless, for the method that is presented in this book.

7

Copy number variations

7.1 Introduction

In the past decade or so, the story of biology in general, and genetics in particular, has been the story of the development of high throughput technology for reading biological sequences, primarily DNA and RNA sequences. At least, this has been my impression. Immediately after a new platform for RNA expression was announced everybody hurried to use that platform and publish a list of genes that show elevated expression under some specified conditions. Likewise, a purchase of a new genotyping machine resulted in a rush of publications that associated that gene or an other with the phenotype close to the researcher's heart. I am not complaining. They all need statisticians to help them analyze their data and this is a way for people like myself to make a living.

And statisticians, or more generally data-management and data-analysis people, are definitely needed. The data that is generated by these technologies is characterized by two features: quantity and error. The statement about quantity is clear. The number of measurements that are produced by a single application of a platform, a number that was in the order of several hundreds only a decade ago and was considered revolutionary, is now reaching the order of several millions and the technologies continue to develop. Handling this multitude of data is a profession.

But the issue of systematic and random error is as important and should be recognized and dealt with. What enabled the revolution was the development of technologies for performing experiments in parallel and in a very small space. In principle, the same type of experiments that were originally conducted individually, each in its own test tube, is now done on tiny bids or on the surface of a chip, all at once and using the same chemical solution. Sometimes my children

---

Extremes in Random Fields: A Theory and its Applications, First Edition. Benjamin Yakir.

have difficulty not interfering in each other's affairs when they are together at home. They are only 4 and the apartment is of a reasonable size. Now condense it and multiply it by a million . . . .

Still, you are not reading this text to learn about my domestic affairs but in order to relate the theory of random fields to real life problems. In this chapter we will deal with a specific application that arises in the context of biological sequence data and see the relevance of random fields for handling the issue of random noise.

The problem is the detection of inherited DNA copy number variants (CNVs). CNVs are gains and losses of (usually very small) segments of the DNA molecule in a chromosome, and comprise an important class of genetic variation in human populations. The modern approach for detecting CNVs involves measurements of genetic markers as probes.

Genetic markers are mapped loci on the genome where genetic measurements can be made. In the simplest form, the measurement involves an intensity that reflects the copy number at that locus. The standard copy number of 2 is associated with some average intensity level. Variation of this number, say to 0, 1, 3, or any other number is associated with a change in that average intensity: a decrease if the copy number is below 2 and an increase if it is more than 2. A huge number of such probes, each reflecting a distinct position on the genome, are measured in parallel for the same individual.

As promised, these measurements involve errors. Consequently, it is difficult to determine on the basis of a single measurement the presence of a CNV at a given locus or the absence thereof. A partial solution is to use repeated measurements. However, applying the platform is costly. Applying it more than once on each subject, especially when the experiment involves a sample of subjects, is currently still prohibitively expensive. A more economical solution is to average measurements in neighboring loci in order to produce a typical intensity for a genomic interval. This approach has the advantage of increasing the strength of a signal if the same CNV extends over several markers. On the other hand, the signal from a very short CNV may be diluted by including in the average intensities from markers that do not belong to the CNV. Nonetheless, this is the method of choice.

The resulting scanning statistic, the statistic that is produced by a standardized average of the intensities from markers in a given genomic interval, has approximately a normal distribution. This is the case, in particular, if the data analyst did a proper job in the preliminary preparation and normalization of the data. This preliminary work is essential, and may involve the identification and removal of systematic errors, the handling of unwanted correlations, and the application of normalizing transformations to the data to make their distribution closer to the normal. Failing to do this step may produce results that reflect flaws in the technology instead of biological phenomena of interest.

Assuming the normality and the independence of the per-marker intensities, an assumption we henceforth make, we get that the basic scanning statistic has a normal distribution. Moreover, if we consider the specifications of the interval, say the location of the center t and the length of the interval h, as parameters we obtain that the resulting random field is Gaussian. Specifically, under the null assumption of the absence of CNVs in an entire section of the chromosome, we get that the distribution of the field is practically identical to the distribution that we obtained in Chapter 2 in the context of a scanning statistic that uses the interval $[-0.5, 0.5]$ as a kernel. In particular, we may use (4.2) to approximate the probability of a false detection in the interval. Alternatively, we may use the discrete version of this approximation, a version that uses the normal random walk overshoot function ν, if we feel that markers are not dense enough to justify the continuous field approximation.

The field in this case of scanning for CNVs in a single individual is Gaussian and as such may be handled by traditional techniques such as the double-sum method. This example hardly serves as a justification for our method. Instead, we will present and analyze an alternative that is designed to use the data from a sample of subjects in order to identify inherited CNVs that are present in the population. The approach that we consider was proposed in [19] and the statistical properties of the approach were discussed in [20, 21].

The detection of CNVs is relevant in diagnostics of cancer. Part of the process that produces abnormal cancer cells results from pathogenic variations in the genetic material of somatic cells, an important part of which is in the form of CNVs. Such variations are very irregular and are specific to the individual tumor. For such irregular behavior a method that concentrates on the data from that individual is more appropriate.

However, CNVs are also a relatively benign part of the normal genetic variation and can be passed on by gamete cells from parents to offspring. Such variations may spread in the population and be present in a noticeable fraction of a sample from that population. Consequently, one may propose to accumulate data on the CNV in a given genomic interval from the sub-sample that carries that CNV as yet another method to overcome random noise and enhance the chances of successful detection.

In the next section we formalize a statistical model for the data and describe the proposed method for accumulating data from sub-samples. The result is a new form of scanning statistic that is again parameterized by t and h but is not a Gaussian field. In Section 7.3 we analyze the statistical properties of the scanning statistic via the application of the measure-transformation technique.

7.2   The statistical model

We assume that data from different subjects are independent of each other. We further assume that the data from each subject were normalized to produce measurements which, in the absence of a signal, are independent standard normal random variables. The observed data may be arranged as a two-dimensional array $\{X_{it} : 1 \leq i \leq n, t_0 < t \leq t_1\}$, where X_{it} is the data point for the ith subject at location t. The number n is the total number of subjects, and $t_1 - t_0$ is the total number of markers. In current genome-wide scanning studies, n may range from several scores to thousands and the number of markers is several hundreds of thousands to about one or two million. An artificial example of a data matrix is presented in Figure 7.1. This matrix corresponds to genetic data from 10 subjects over 100 genomic markers. Markers in this artificial example are placed 1 kb apart. Subjects 2, 3, 7, and 10 in this example have elevated levels of expected expression in marker positions 20 to 30.

Given a genomic interval $[t - h/2, t + h/2]$, parameterized by $\theta = (t, h)$, we compute a summary statistic for the evidence from subject i regarding the presence of a CNV in the interval:

$$Z_{i,\theta} = \frac{1}{\sqrt{h+1}} \sum_{j=t-h/2}^{t+h/2} X_{ij} \, .$$

This statistic would have been the scanning statistic had we considered only the data from subject i. The statistic tends to obtain values in an interval about the origin if the subject does not carry a CNV, it has the standard normal distribution in this case, or else it tends to have a more extreme value – positive or negative – if it does carry a CNV.

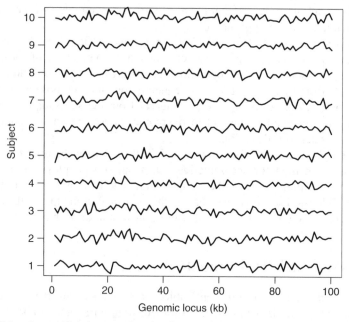

Figure 7.1   An artificial example of a data matrix. The raw data for 10 subjects over 100 genomic markers are presented, with 1 marker set every 1 kb. Subjects 2, 3, 7, and 10 have elevated levels of the expectation for the markers in positions 20 to 30.

The question is how to accumulate the data from a sample of subjects of which only some, if any all, are carriers of a CNV? Taking a sum may not be a good idea, especially if the carriers of the CNV are a small minority in the sample, since that will dilute the signal and make it harder to detect. Even worse, since a CNV can be expressed for some in the form of an increase in the intensity and for others in the form of a decrease in the intensity, the signal may be canceled out by the summation. This problem of opposite-sign cancelation may be solved by taking the absolute value of $Z_{i,\theta}$ or squaring it before the summation, still leaving the dilution of the signal as a concern.

An alternative would be to sum the statistics, or their absolute values, only for the subjects in the sub-sample that carries a CNV. Unfortunately, we do not know who these subjects are. Yet, this is not a silly proposal since we do have data on each subject in the form of $Z_{i,\theta}$. Thus we can identify an ad hoc sub-group of patients with relatively elevated values of $|Z_{i,\theta}|$ and form the sum for these subjects, and for these subjects only. A generalization of the same idea may apply a function $g(z)$, a function that produces relatively small values when the absolute value z is not sufficiently large but produces high values when the absolute value of z is large. Such a function can be used on order to construct a scanning statistic:

$$Y_\theta = \sum_{i=1}^{n} g(Z_{i\theta}) . \tag{7.1}$$

If that scanning statistic is applied and detections declared whenever a threshold y is crossed then the probability $P(\max_{\theta \in T} Y_\theta \geq y)$ reflects the rate of false detections.

There is flexibility in the selection of g. Taking g to be the identity function will produce the sum of the subject-specific scanning statistics. This approach was rejected as a bad one. Still, for this selection of g we get that the sum has a Gaussian distribution and we already have an approximation for the rate of false alarms for this case.

Taking the sum of squares of the subject-specific scanning statistics corresponds to choosing $g(z) = z^2$. The distribution of the scanning statistic will no longer be normal. Instead it will be the chi-square distribution on n degrees of freedom. It requires work, but one may squeeze the theory of Gaussian fields to allow results for this chi-square field. However, for other selections of g the field will not be Gaussian nor will it be linked to a Gaussian field other than the general statement that with the increase in n the field converges to a Gaussian field.

For example, the idea of preselecting a sub-sample for producing the statistic may be obtained with a threshold function that is equal to zero for values less than the threshold and equal to the identity for values larger than the threshold. Such a function can be applied to the absolute value of the square of $Z_{i\theta}$. A similar approach, which is motivated by ideas stemming from a mixture of distributions, is to use the function $g(z) = \log(1 - p + p \cdot \exp\{z^2/2\})$, where p is a parameter that represents the frequency of the CNV in the population. This function is a smooth function, a fact that facilitates the analysis that we will carry out, although the analysis can be conducted with slightly more effort for the noncontinuous threshold function.

For illustration we computed the scanning statistic that results from the application of the function $g(z) = \log(1 - p + p \cdot \exp\{z^2/2\})$, for $p = 0.1$. We used $h = 14$ and computed the standardized interval specific summary statistics for each subject and each location t. Locations near the edge involved truncated intervals. The statistic Y was computed for each location via the application of the function to the subject-specific summary statistics that are associated with the locations and the summation of the 10 resulting values. The plot of the scanning statistic Y_θ is presented in Figure 7.2.

The resulting random field Y_θ, for both the threshold function or the mixture-type function, is not Gaussian nor is it directly related to a random field in a way that allows application of traditional methods for Gaussian fields. Admittedly, the limit distribution of the field when n diverges to infinity is Gaussian. However, we may remind the reader of the dangers in applying Gaussian approximations to the extreme tail of a non-Gaussian field. In the current setting, if the threshold in the threshold function is relatively large or if the parameter p in the mixture-type function is small then the distribution of Y_θ will be severely skewed.

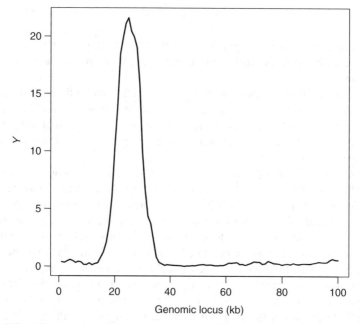

Figure 7.2   The computed scanning statistic. The function $g(z) = \log(1 - p + p \cdot \exp\{z^2/2\})$, for $p = 0.1$, was applied to the standardized interval summary statistics computed from the raw data that is presented in Figure 7.1. The summary statistics were computed for $h = 14$. The statistic Y was obtained by summation of the subject-specific outcomes across all 10 subjects.

A normal approximation, especially for values of n not extremely large, will be questionable.

The inability to apply the standard theory of maxima in Gaussian fields is exactly the property that we cherish. The standard Gaussian techniques do not work but the measure-transformation approach does. In the next section we apply this approach to the random field Y_θ in order to obtain approximations of the rate of false discoveries. For convenience, we will make the assumption that the function g is differentiable and leave the case where it is not as one of the projects/excercises for the reader.

7.3   Analysis of statistical properties

We derive the asymptotic expansion of the probability $P(\max_{\theta \in T} Y_\theta \geq y)$, for $Y_\theta = \sum_{i=1}^{n} g(Z_{i,\theta})$, as a function of the threshold y and the sample size n. The parameter set is $T = [t_0, t_1] \times [h_0, h_1]$. In the expansion we will let both y and n grow to infinity. As you may read between the lines below, the analysis can be made rigorous also in the case where n is kept fixed. However, the large sample asymptotic formulae are simpler and the proof, in light of Theorem 5.3, is more streamlined.

7.3.1   The alternative distribution

The first decision to make is which alternative distribution to use. There is no natural built-in alternative. Consequently, we choose to use the omnibus technique of exponential tilting. This technique can also help us identify the large deviation factor.

It is helpful to start by tilting the distribution of a statistic for a single subject and then extend to the entire sample. Define the log-moment generating function $\psi(\xi) = \log E \exp\{\xi g(Z)\}$, where Z is a standard normal random variable. Observe that $Z_{i,\theta}$ has the standard normal distribution for all i and θ. It follows that $\ell_{i,\theta} = \xi g(Z_{i,\theta}) - \psi(\xi)$ is a log-likelihood that is associated with the ith sample. One may extend this basic construction in order to produce a tilted distribution for the entire sample by taking the sum and considering the likelihood ratio $\ell_\theta = \sum_{i=1}^{n} \ell_{i,\theta}$. Under the tilted distribution the samples are still independent of each other, each of them tilted according to the parameter ξ. The statistic Y_θ, a sum of independent elements from the tilted distribution, inherits this distribution.

The large deviation properties of the marginal probability $P(Y_\theta \geq y)$, and as a byproduct, the value of the parameter ξ that will be used may be obtained from the examination of the log-likelihood function of Y_θ, which is $n\psi(\xi)$. The expectation of Y_θ under the tilted distribution is $n\psi'(\xi)$, where ψ' is the derivative of ψ. The value of ξ that is associated with the large deviation rate is the value that equates the expectation with the threshold:

$$n\psi'(\xi) = y \quad \Rightarrow \quad \xi = \xi(y, n) = [\psi']^{-1}(y/n) , \tag{7.2}$$

which we use henceforth. With this selection of parameter ξ the large deviation factor that will be produced is $\exp\{n[\psi(\xi) - \xi\psi'(\xi)]\}$. Since the distribution of the statistic Y_θ is the same for all θ we will get the same factor all across the parameter set.

7.3.2   Preliminary localization and approximation

A preparation step that precedes the application of the likelihood ratio identity is the initial localization and the approximation of the original parameter set by a discrete subset. In the current example this preliminary step is not needed. The marginal distributions of the random field are identical to each other for all values of the parameter, hence there is no subset of parameters that is associated with a larger large deviation factor. Likewise, the parameter space is discrete to begin with, no further action is required. However, if it is judged that the density of genetic markers, though discrete, is not consistent with the asymptotic derivation we intend to perform then our current decision not to apply this step may be revised. Furthermore, as a premonition we do want to maintain the flexibility to set the appropriate range of values of the parameter set, namely the selection of t_0, t_1, h_0, and h_1 as a function of y of n, to assure that the asymptotic derivation that we produce is meaningful. This freedom of choice and the decision regarding the approximation of the parameter set by a subset are closely related and reflect the larger issue of determining appropriate scalings.

7.3.3   Measure transformation

We turn to the transformation of the measure. We produced above an alternative distribution associated with a specific parameter value θ and the random field element Y_θ that is connected to it. This distribution, defined for a specific element, needs to be extended to the entire random field. We do so by defining the alternative distribution for the field in a way that retains ℓ_θ as the likelihood ratio. Hence, if dP is the joint density of all the elements in the field then we define the alternative distribution using the density $dP_\theta = e^{\ell_\theta} dP$. It is worthwhile to pay attention to the fact that the property of independence between subject-specific fields that was in the original distribution is still intact under the tilted distribution. Specifically, if the joint null distribution of the fields is characterized by the product $\prod_{i=1}^n [dP_i]$, with dP the n-fold convolution of the dP_i densities, then the joint alternative distribution of the subject-specific fields is $\prod_{i=1}^n [e^{\ell_{i,\theta}} dP_i]$. The distribution $dP_\theta = e^{\ell_\theta} dP$ is the n-fold convolution of the densities in the square brackets.

With the given alternative distributions the likelihood ratio identity produces:

$$P\left(\max_{\theta \in T} Y_\theta \geq y\right) = \sum_{\theta \in T} E_\theta \left(\frac{1}{\sum_{\vartheta \in T} e^{\ell_\vartheta}}; \max_{\vartheta \in T} Y_\vartheta \geq y\right).$$

The log-moment generating function $n\psi(\xi)$ that appears in each of the log-likelihoods is the same for all values of ϑ. Consequently, it follws that the local field, the collection $\{\ell_\vartheta - \ell_\theta\}$, is composed of elements of the form:

$$\xi(Y_\vartheta - Y_\theta) = \sum_{i=1}^{n} \xi[g(Z_{i,\vartheta}) - g(Z_{i,\theta})] .$$

After the usual rearrangement of terms in each of the expectations that were produced by the likelihood ratio identity these expectations become:

$$E_\theta \left(\frac{1}{\sum_{\vartheta \in T} e^{\ell_\vartheta}} ; \max_{\vartheta \in T} Y_\vartheta \geq y \right) = e^{n[\psi(\xi) - \xi \psi'(\xi)]} E_\theta \left(\frac{M_\theta}{S_\theta} e^{-[\tilde{\ell}_\theta + m_\theta]} ; \tilde{\ell}_\theta + m_\theta \geq 0 \right),$$

where

$$\tilde{\ell}_\theta = \sum_{i=1}^{n} \xi[g(Z_{i,\theta}) - \psi'(\xi)] ,$$

$$S_\theta = \sum_{\vartheta \in T} \exp \left\{ \sum_{i=1}^{n} \xi[g(Z_{i,\vartheta}) - g(Z_{i,\theta})] \right\} ,$$

and

$$M_\theta = \max_{\vartheta \in T} \exp \left\{ \sum_{i=1}^{n} \xi[g(Z_{i,\vartheta}) - g(Z_{i,\theta})] \right\} , \qquad m_\theta = \log M_\theta .$$

7.3.4   The localization theorem and the local limit theorem

We intend to apply the localization theorem, Theorem 5.2, to the expectations that are produced by the likelihood ratio identity with $\kappa = n$, a parameter proportional to the variance of $\tilde{\ell}_\theta$. The application involves determining a local σ-algebra $\hat{\mathcal{F}}_n$ and checking the five conditions of the theorem. For the determination of the local field it is convenient to temporarily change the parametrization to $\theta = (\theta_1, \theta_2)$, with $\theta_1 = t - h/2$ and $\theta_2 = t + h/2$. With this in mind, we set $\hat{\mathcal{F}}_n = \sigma \left\{ \sum_{i=1}^{n} \xi[g(Z_{i,\vartheta}) - g(Z_{i,\theta})] : |\vartheta_j - \theta_j| \leq \tau, j = 1, 2 \right\}$, for some τ. Of the five conditions, Condition I* is trivial, Condition IV* follows directly from the local limit theorem, Theorem 5.3, and Condition III*, in light of the local limit theorem, is an immediate corollary of the identification of the limit expectation and covariance structure of the Gaussian field that results from taking the limit of the local field, jointly with the global term.

In order to understand the issues involved in the derivation of the limit distribution let us consider in detail the expectation of a component of the local field of the form: $\sum_{i=1}^{n} \xi[g(Z_{i,\vartheta}) - g(Z_{i,\theta})]$.

The expectation of the sum is the sum of expectations, each taken under the alternative distribution for the component in the sum. The statistic $Z_{i,\theta}$ is a sufficient statistic for that alternative distribution. Namely, the conditional distribution, given the statistic, is the same as the conditional distribution under the original null distribution. It follows that the expectation is:

$$n\xi E_\theta([g(Z_{1,\vartheta}) - g(Z_{1,\theta})]) = n\xi E([g(\{1-r^2\}^{\frac{1}{2}}W + rZ) - g(Z)]e^{\xi g(Z) - \psi(\xi)}),$$
(7.3)

where the expectation on the right-hand side is taken with respect to the two independent standard normal random variables and

$$r = \operatorname{Cov}(Z_{1,\vartheta}, Z_{1,\theta}) = \frac{\min\{t + \frac{h}{2}, s + \frac{w}{2}\} - \max\{t - \frac{h}{2}, s - \frac{w}{2}\}}{\sqrt{hw}}.$$

The representation results from the regression of $Z_{1,\vartheta}$ on $Z_{1,\theta}$, with W in (7.3) being the standardized residual of the regression.

It is of interest to see under what conditions the expectation would converge to a constant. The assumption that $n \to \infty$ implies that either $\xi \to 0$, $r \to 1$, or both limits are taking place at the same time. The first type of limit corresponds to local alternatives, i.e., the transformation to a measure that differs only slightly from the null distribution and will imply the relation:

$$y = n\psi'(0) + [n\psi''(0)]^{\frac{1}{2}}z \quad \Rightarrow \quad \xi \sim z[n\psi''(0)]^{-\frac{1}{2}},$$

for $z^2/n \to 0$. The second limit corresponds to a high correlation between the statistics. If ξ is kept fixed then $h + 1$, the number of loci involved in the formation of the statistic $Z_{1,\theta}$, should be proportional to n. The requirement of an analysis that combines both types of convergence is that h/z^2 goes to a positive constant.

Which road to take is of course a matter of personal taste regarding appropriate scaling and may depend on the characteristics of the scientific problem as they are understood by the person that develops the approximation. Below we consider the case where ξ is kept fix, so the ratio y/n is converging to a constant that differs from $\psi'(0)$.

After making the decision regarding scaling let us try to produce justifications for the choice. The best justification, in the context of this book, is the fact that the resulting asymptotic derivation is as far as it can be from the asymptotic approximation produced by a naïve application of the formulae for the maximum of a Gaussian field to the limit Gaussian distribution of the standardized statistics $[Y_\theta - n\psi'(0)]/[n\psi''(0)]^{\frac{1}{2}}$. A second justification can be made on a more scientific basis. The level of the required threshold y is a reflection of the effective number of elements in the field, which in our case is of the order of magnitude of the product of the number of genetic markers times the sample size, since we will take h to be proportional to n. The number of markers is several orders of magnitude larger than the sample size n, and that should be reflected in a very large threshold y if an effective control on the rate of false detection of CNVs is intended.

With the choice we made, let us return to the expectation of the element of the local field that is given in (7.3). We are in a situation where r is converging to 1 and g is a smooth function. Define $V_r = V = \{1 - r^2\}^{\frac{1}{2}} W + (r - 1)Z$ and take a second-order Taylor expansion of the function g about Z to get:

$$
n\xi E_\theta([g(Z_{1,\vartheta}) - g(Z_{1,\theta})]) = n\xi E\left([g(Z + V) - g(Z)]\, e^{\xi g(Z) - \psi(\xi)}\right)
$$

$$
= n\xi E\left(\left[g'(Z)V + \frac{1}{2}g''(Z)V^2\right] e^{\xi g(Z) - \psi(\xi)}\right) + O([1 - r]^3 n)
$$

$$
= -(1 - r)n\xi E\left([Zg'(Z) - g''(Z)]e^{\xi g(Z) - \psi(\xi)}\right) + O([1 - r]^2 n),
$$

that follows from the independence between the standard normal variables W and Z and the fact that $1 - r^2 \approx 2(1 - r)$. Integrating by parts we get that

$$
\int g''(z)e^{\xi g(z) - \frac{1}{2}z^2}dz = -\int (\xi[g'(z)]^2 - zg'(z))e^{\xi g(z) - \frac{1}{2}z^2}dz ,
$$

which implies that:

$$
n\xi E_\theta([g(Z_{1,\vartheta}) - g(Z_{1,\theta})])
$$
$$
\approx -n[1 - \text{Cov}(Z_{1,\vartheta}, Z_{1,\theta})]\xi^2 E([g'(Z)]^2 e^{\xi g(Z) - \psi(\xi)}) . \qquad (7.4)
$$

For the variance of an increment of the local field we may use independence of the random fields to conclude that the variance of the sum is n times the variance of an increment of a single random local field. Here it is sufficient to take a first-order Taylor expansion of the function g:

$$
n\text{Var}_\theta(\xi[g(Z_{1,\vartheta}) - g(Z_{1,\theta})]) = n\xi^2 \text{Var}\left([g(Z + V) - g(Z)] e^{\xi g(Z) - \psi(\xi)}\right)
$$

$$
= n\xi^2 \text{Var}\left([g'(Z)V] e^{\xi g(Z) - \psi(\xi)}\right) + O([1 - r]^2 n)
$$

$$
= (1 - r^2)n\xi^2 E\left([g'(Z)]^2 e^{\xi g(Z) - \psi(\xi)}\right) + O([1 - r]^{1.5} n)
$$

$$
\approx -2nE_\theta(\xi[g(Z_{1,\vartheta}) - g(Z_{1,\theta})]) ,
$$

with the passage from the second to the third line being partially justified by the computation of the variance via the conditioning on the value of the random variable Z. We obtained that the asymptotic variance of an increment is twice the absolute value of the asymptotic expectation of the increment, a characteristic feature of a Gaussian log-likelihood ratio when testing for a nonzero mean vector. Specifically in this case, an asymptotic expansion of the covariance between two increments of the local field, which can be conducted by considering again a first-order Taylor expression, demonstrates that the limit distribution of the local random field is that of a two-sided random walk of Gaussian likelihood ratios for testing the hypothesis that the expectation of the increments is some fixed value. We omit the details.

For the asymptotic expression in Condition III* we may use the fact that:

$$1 - \text{Cov}(Z_\theta, Z_\vartheta) \approx \frac{1}{2h}(\vartheta_1 - \theta_1) + \frac{1}{2h}(\vartheta_2 - \theta_2) \,,$$

to obtain expressions that are similar to the expressions that emerged in the similar situation of a scanning statistic with an indicator of an interval as a kernel, an example that was among the basic examples of Chapter 2. Specifically, in the current case and in the context of Condition III*, we get that $E[\mathcal{M}/\mathcal{S}] = \{I(\theta)v([2I(\theta)]^{\frac{1}{2}})\}^2$, where v is the overshoot function of a Gaussian likelihood ratio random walk and:

$$I(\theta) = \frac{n}{h}\frac{\xi^2}{2}E_\theta\left[\{g'(Z)\}^2\right] = \frac{n}{h}\frac{\xi^2}{2}\int [g'(z)]^2 e^{\xi g(z) - \psi(\xi)}\phi(z)dz = \frac{n}{h}\cdot \iota(\xi) \tag{7.5}$$

is the expectation of an increment of the random walk.

The expectation of the global term $\tilde{\ell}_n$ is $\mu = 0$ by construction. The variance of the global term is:

$$\text{Var}_\theta(\tilde{\ell}_n) = n\xi^2\text{Var}_\theta(g(Z_{1,\theta})) = n\xi^2\psi''(\xi) = n\sigma^2 \,.$$

Jumping our guns, we may predict that as a result of the integration step, if we take $h_0 = n\eta_0$ and $h_1 = n\eta_1$, and if we change to the variable $w = h/n$, then we will obtain the approximation:

$$P\left(\max_{\theta \in T} Y_\theta \geq y\right) \approx \frac{e^{n[\psi(\xi) - \xi\psi'(\xi)]}(t_1 - t_2)n}{\{2\pi n\xi^2\psi''(\xi)\}^{-\frac{1}{2}}}\int_{\eta_0}^{\eta_1} \{(\iota(\xi)/w)v([2\iota(\xi)/w]^{\frac{1}{2}})\}^2 dw \,. \tag{7.6}$$

However, before reaching this conclusion we still need to finish validating the appropriateness of Theorem 5.3 for the current problem and to check the two remaining conditions of Theorem 5.2.

In order to complete the checking of the conditions for Theorem 5.3 we want to show that the global term and an increment of the random field are asymptotically uncorrelated. Taking a first-order Taylor expansion we obtain:

$$n\xi^2\text{Cov}_\theta(g(Z_{1,\theta}), g(Z_{1,\vartheta}) - g(Z_{1,\theta}))$$

$$\approx n\xi^2 E\left([g(Z) - \psi'(\xi)][g'(Z)V]e^{\xi g(Z) - \psi(\xi)}\right).$$

The covariance between two random variables is the expectation of the product of the centered random variables. It is enough to center one of the two variables. We arrived at the approximation by considering the centered global term and a first-order approximation of the increment of the local field. We used once more independence between subject-specific fields to represent the covariance as a sum of covariances.

The resulting approximation of the covariance is of the order of a constant for $V = \sqrt{1 - r^2}\, W - (1 - r)Z$, since the term associated with W makes a zero contribution and $n(1 - r)$ converges to a constant. Consequently, the correlation between the global term and the local field is of order $n^{-\frac{1}{2}}$ and converges to zero.

Regarding the rest of the unchecked conditions of Theorem 5.2, they can be verified using methods that are not unlike the methods that were used in the context of Gaussian scanning statistics.

7.3.5 Checking Condition V*

For example let us look at Condition V*. For the same reasons that were given for scanning statistics it is sufficient to bound the right tail of M_θ. A trivial bound can be provided by taking:

$$P_\theta(M_\theta > e^x) = P_\theta \left(\max_{\vartheta \in T} \exp \left\{ \sum_{i=1}^{n} \xi[g(Z_{i,\vartheta}) - g(Z_{i,\theta})] \right\} > e^x \right)$$

$$\leq |T| P_\theta \left(\exp \left\{ \sum_{i=1}^{n} \xi[g(Z_{i,\vartheta}) - g(Z_{i,\theta})] \right\} > e^x \right) \leq |T| e^{-x} ,$$

by the fact that $\sum_{i=1}^{n} \xi[g(Z_{i,\vartheta}) - g(Z_{i,\theta})] = \ell_\vartheta - \ell_\theta$ is a log-likelihood ratio under the P_θ distribution and the application of the Markov inequality.

The derivation of (7.6) requires that $t_1 - t_2 \gg h_1 - h_0 = (\eta_1 - \eta_2)n$. Consequently, $|T|$ should grow faster than n^2 but, otherwise, any polynomial rate will do. Consequently, $x = (c + 0.5) \log n$ is sufficient for c larger than the polynomial order of $|T|$.

7.3.6 Checking Condition II*

Condition II* is the last on the list. This condition deals with the probability $P_\theta(A_{\mathrm{II}}^c \cap \{\tilde{\ell}_\theta + \hat{m}_\theta \in x + (0, \delta]\})$, for a truncated $\hat{m}_\theta$ and for x in a relatively narrow range. We may apply the same approach that was used for a Gaussian scanning statistic:

$$P_\theta(A_{\mathrm{II}}^c \cap \{\tilde{\ell}_\theta + \hat{m}_\theta \in x + (0, \delta]\})$$

$$\leq P_\theta(A_{\mathrm{II}}^c \cap \{-m < X_\theta \leq \delta\}) + P_\theta(\{X_\theta + \hat{m}_\theta \in (0, \delta]\} \cap \{\hat{m}_\theta > m\}) ,$$

for $X_\theta = \tilde{\ell}_\theta - x$ and a finite m.

If m is large enough to make the probability of the event $A = \{\hat{m}_\theta > m\}$ small then we will have by Theorem 5.3 that the last probability, multiplied by $\sqrt{n}$, converges to a quantity no larger than $P(A)/\sqrt{2\pi}$.

For the first of the two probabilities in the bound we consider the sum of probabilities:

$$\sum_{\tau < \|\vartheta - \theta\| \le \log n} P_\theta(\{\ell_\vartheta - \ell_\theta \ge \log(\epsilon p_\vartheta) \cap \{-m < X_\theta \le \delta\}$$

$$+ \sum_{\|\vartheta - \theta\| > \log n} P_\theta(\ell_\vartheta - \ell_\theta \ge \log(\epsilon p_\vartheta)) .$$

This statement follows from the fact that the event A_{II}^c is included in a union. For elements in the sum that are associated with remote parameters we ignore the intersection with the event that involves the global term.

We use log-moment generating functions in order to bound the probabilities, either for small values of $\|\vartheta - \theta\|$ or for larger values. In both cases we consider the conditional log-moment generating function, given the values of $\{Z_{i,\theta}\}$, evaluated at $1/2$. Specifically we are interested in terms of the form:

$$\psi_r(\xi/2|Z) = \log E(\exp\{(\xi/2)[g(V_r + Z) - g(Z)]\}|Z)$$

for $V_r = (1 - r^2)^{\frac{1}{2}} W - (1 - r)Z$.

The main concern is for values of θ for which $1 - r$ is small. For such values we expand the function g, taking a two-term Taylor approximation:

$$\psi_r(\xi/2|Z) = \log E(\exp\{(\xi/2)[g'(Z)V_r + 0.5g''(Z)V_r^2]\}|Z) + R .$$

If the function g has a locally bounded third derivative then the error term R is $O_p([1 - r]^{\frac{3}{2}})$ and may be ignored, even after multiplication by n, for r such that $(1 - r)n(\log n)^{-1} \to 0$. For larger r it may not produce a vanishing term after multiplication by n, but it cannot significantly change the effect of the leading term.

Keeping track only of terms that make a nonvanishing contribution to the conditional moment generating function yields:

$$g'(Z)V_r + \frac{1}{2}g''(Z)V_r^2 \approx (1 - r^2)^{\frac{1}{2}}g'(Z)W + (1 - r)g''(Z)W^2 - (1 - r)Zg'(Z) .$$

This term should be multiplied by $\xi/2$, exponentiated, and integrated with respect to the density of W. Notice that the term associated with W^2 and the term $-W^2/2$ that appears in the exponent of the density can be combined, changing in effect the variance, which is 1 in the original density, to $[1 - \xi(1 - r)g''(Z)]^{-1}$. The expectation is still equal to zero. The outcome from the integration of the approximation is:

$$\psi_r(\xi/2|Z) \approx -(1 - r)(\xi/2)Zg'(Z) - \frac{1}{2}\log(1 - \xi(1 - r)g''(Z))$$

$$+ \frac{\xi^2}{8}(1 - r^2)[g'(Z)]^2(1 - \xi(1 - r)g''(Z))$$

$$\approx -(1-r)(\xi/2)\{Zg'(Z) - g''(Z) - (\xi/2)[g'(Z)]^2\}\ .$$

Consider the approximation of the conditional log-moment generating function as a function of Z. Recall that $E_\theta[Zg'(Z) - g''(Z)] = \xi E_\theta[(g'(Z))^2]$. Consequently, the expectation of the approximation is $-(1-r)(\xi/2)^2 E_\theta[(g'(Z))^2]$. The variance is proportional to $(1-r)^2$, i.e., of a much smaller order.

Define $U_\vartheta = \sum_{i=1}^n \{\psi_r(\xi/2|Z_{i,\theta}) - E_\theta \psi_r(\xi/2|Z_{i,\theta})\}$, with $r = \text{Cov}(Z_{i,\vartheta}, Z_{i,\theta})$. The bound on the probability will follow from the relation:

$$P(\ell_\vartheta - \ell_\theta \geq \log(\epsilon p_\vartheta)|\{Z_{i,\theta}\}) \leq \exp\{n E_\theta \psi_r(\xi/2|Z_{1,\theta})$$
$$- (1/2)\log(\epsilon p_\vartheta) + U_\vartheta\}\ .$$

Consequently, Condition II* is a result of the examination of the bound:

$$\frac{c(m+\delta)}{\sqrt{n}} \times \sum_{\tau < \|\vartheta - \theta\| < \log n} \frac{e^{-(n/2)E_\theta \psi_r(\xi/2|Z_{1,\theta})}}{\sqrt{\epsilon p_\vartheta}}$$

$$+ \sum_{\log n < \|\vartheta - \theta\|} \frac{e^{-(n/2)E_\theta \psi_r(\xi/2|Z_{1,\theta})}}{\sqrt{\epsilon p_\vartheta}}$$

$$+ \sum_{\tau < \|\vartheta - \theta\|} P_\theta(U_\vartheta > (n/2)E_\theta \psi_r(\xi/2|Z_{1,\theta}))\ .$$

The constant c that appears in the term that multiplies the first sum is produced by the application of the Berry–Esseen theorem to X_θ, assuming that a third moment for the increments exist. Selecting $p_\vartheta \propto \exp\{-\varepsilon(n/2)E_\theta \psi_r(\xi/2|Z_{1,\theta})$, for a small enough $\varepsilon > 0$, will assure that the first two sums are $o(n^{-\frac{1}{2}})$.

In order to complete the story we need to deal with the probabilities. Restricting ourselves to the case where $E_\theta\{\psi_r(\xi/2|Z_{i,\theta}) - E_\theta \psi_r(\xi/2|Z_{i,\theta})\}^4$, the centered fourth moment, is finite and asymptotic to $(\|\vartheta - \theta\|/n)^4$ we may use the bound:

$$P_\theta(U_\vartheta > (n/2)E_\theta \psi_r(\xi/2|Z_{1,\theta})) \leq \frac{E_\theta(U_\vartheta^4)}{\{(n/2)E_\theta \psi_r(\xi/2|Z_{1,\theta})\}^4} \approx \frac{c}{n^2}\|\vartheta - \theta\|^{-4}\ ,$$

for some limit constant c. But, $\sum_{\vartheta \in T}\|\vartheta - \theta\|^{-4} < \infty$, uniformly in n. Therefore, the contribution of the sum of probabilities is negligible.

This completes the validation of the conditions for Theorem 5.2 and establishes the approximation (7.6). It goes almost without saying that this approximation can be generalized to the case where the probability does not converge to zero with the aid of a Poisson approximation. The generalization applies the function $f(\lambda) = 1 - \exp\{-\lambda\}$ to the term that is given on the right-hand side of (7.6). Interestingly enough, the Poisson approximation has further implications in this example which we would like to discuss.

7.4   The false discovery rate

The context in which this book is written is controlling the error rate. The computations are conducted under the null distribution where signals are completely absent. The probability that is being computed is associated with wrongly declaring the presence of a signal when we should have not done so. The motivation for wanting to verify that the probability of falsely rejecting the null hypothesis is small is the worry that the consequences for such an error may be severe and should be avoided as much as reasonable.

There are other situations, among them the detection of CNVs, where making an error as mentioned is not too catastrophic. Usually, one may apply other laboratory techniques, that are albeit more costly per locus, in order to validate that an identified CNV is genuine. Hence, one may consider the application of the high throughput technology as a screening step which is used in order to eliminate the parts of the genome that do not carry such variations in order to concentrate subsequent efforts on the parts that most likely express the variations. Still, one may not want the remaining regions to be polluted by too many false positives. A reasonable approach to address this concern is to use a statistical method that controls the rate of false detection.

In the statistical literature such methods fall under the acronym FDR, which stands for false discovery rate. The standard technique deals with a setting that involves independent random variables and does not fit the situation of a random field with dependence between elements. However, there are methods that can be used in the given context.

One such method will call for a more careful definition of what constitutes a discovery. In the case of screening for CNVs, for example, one may relate a region in which the threshold was crossed to a discovery, which may be true or false. Still, the exact meaning of the notion of a region may need to be specified but we will not worry about it here. The FDR is the expected ratio of false discoveries to the total number of discoveries. The method makes sure that this expected ratio is not above some pre-specified level.

Notice that we no longer deal with a distribution which is generated purely from the null distribution. Currently, we are in a situation where some parts of the parameter set are associated with a genuine signal whereas in other parts there is only random noise. The computations that we carried out were not designed to work in such a setting. However, if we restrict the computations to the part of the parameter space devoid of true signals, presumably the greater part of the parameter set, then the results that we obtained are still valid.

In order for the above-mentioned method for controlling the FDR to work, two conditions should be met. The first is that the number of true discoveries is asymptotically independent of the number of false discoveries. The other condition is that the asymptotic distribution of the number of false discoveries is Poisson. There is no other restriction on the distributions involved. The results that we obtained are relevant for the justification of the second condition.

Indeed, under a reasonable association of the count of discoveries with the number of connected excursion sets that are associated with a given threshold, we will get that this count follows, approximately, the Poisson distribution. The Poisson probability of the count k is given by the application of the function $f_k(\lambda) = e^{-\lambda}\lambda^k/k!$ to the expression on the right-hand side of (7.6). This computed value of λ is used for the application of the method.

The simple method for applying the FDR methodology is discussed in [22]. The proposed methodology associates discoveries with excursion sets that associated with a level y. Let $R = R_y$ be the number of such sets, which is considered as the count of the number of claimed discoveries. The number of false discoveries among them, the number of excursion sets that do not contain a genuine CNV, is denoted by $V = V_y$. The proportion of false discoveries is V/R, and the FDR is $E(V/R; R > 0)$.

Based on the material presented in Chapter 4 we can infer that the asymptotic distribution of V_y is Poisson with rate $\lambda = \lambda_y$ that is derived from the right-hand side of (7.6). For the most part, if genuine CNVs are not so common relative to the scale of the entire genome, then one may expect most of the elevated regions of the random field that contribute to the count V not to occur in the vicinity of an actual CNV. Consequently, the total number of such regions should be asymptotically independent of $R - V$, especially if care is taken not to include in the count elevations that are in the immediate vicinity of a given elevation. Consequently, we are in a situation that is assumed for the FDR methodology.

The methodology itself proposes to use $\lambda_y/(R_y + 1)$ as an unbiased estimator of the FDR that results from the application of the threshold y. Alternatively, one may select a level y to assure a given FDR. Indeed, for α, a pre-specified bound on the value of the FDR, one may choose y to be the smallest threshold that still obeys the relation $R_y \geq \lambda_y/\alpha$. For the resulting threshold the expected proportion of false discoveries is no larger than α.

8

Sequential monitoring of an image

8.1 Introduction

You watch a movie. It is a boring movie. Actually, it is a very boring movie. Nothing happens. All you see are random dots that flash and go. Suddenly, at some point in time, a higher intensity of random points is beginning to take shape at some location on the screen, while at other locations random points still come and go as before. The target is to set an alarm immediately after the emergence of the shape. Randomness may also create misleading shapes and cause false alarms. The monitoring procedure should be tuned to have false alarms only rarely.

In this application we move from a problem in the analysis of biological sequence data, a problem we know a little about, to a problem of image processing, a field in which we lack any such knowledge. Let that not stop us. We will propose a scenario that seems to us to make sense and analyze relevant statistical properties using the tools that we now possess. Connection to reality, or lack thereof, will be of lesser importance. The ability to make meaningful mathematical statements in a not completely trivial setting is our goal.

The setting in which we place our story is sequential change-point detection. The basic situation is similar to the case of industrial quality control that was discussed in Chapter 4. The context there was sequential monitoring of a quality index with an intention to detect deterioration in the quality of the production as soon as possible after deterioration began. The stream of incoming data in that simple example came in the form of periodic measurements. In that simple setting we allowed only two type of distributions, f when the system is in control

Extremes in Random Fields: A Theory and its Applications, First Edition. Benjamin Yakir.
© 2013 by Higher Education Press. All rights reserved. Published 2013 by John Wiley & Sons, Ltd.

and g when it is out of control. The only unknown parameter in the model was the time of change from distribution f to the distribution g.

The specific statistic that was employed as the quality index was the cusum statistic. This statistic is constructed from log-likelihood ratios. After the accumulation of n observations, the log-likelihood statistic for testing a change that was initiated at the kth observation is of the form of a partial sum. The elements in the sum are the observation-specific log-likelihood ratios, extending from the kth observation to the current nth observation. The cusum statistic itself is produced by maximizing the partial sums over the values of k, $1 \leq k \leq n$. The computed sequence of the cusum quality index determines the monitoring procedure's stopping time. Immediately after the process of cusum statistics crosses a threshold x a change is declared to have happened.

A false detection takes place if a change is declared prior to the occurrence of a change. The problem we concerned ourselves with was the controlling of the rate of false detection. This control is obtained by assuring that the expectation of the stopping time is set at a given (high) level. The relevant expectation is computed in the model that involves no change, a model in which all observations come from the distribution f.

The road that led to the approximation of the expectation was a winding one. It involved obtaining the limit distribution of the stopping time and using dominated convergence in order to equate the limit of the expectation of the stopping time with the expectation of the limit distribution. The limit distribution itself turned out to be exponential, with a rate that is determined by the probability of stopping within a relatively short interval of time. The probability of stopping within a short interval was approximated using the measure-transformation technique and the transformation from the local computation of the rate to the global statement regarding the exponential distribution was obtained via the application of a Poisson approximation.

The measure-transformation technique that produced an approximation for the rate benefited from the representation of the event of stopping within a given time interval in terms of an event expressed as an extreme of a random field. The components of the field were the log-likelihood partial sums, parameterized by both endpoints k and n. The resulting parameter space formed a triangle of paired endpoints. The event that the maximum of this field is larger than the threshold x coincides with the event the cusum process goes above the threshold within the given time interval.

The current story involves monitoring a sequence of images that emerge, say, as movie frames. The aim is to detect the appearance of a feature of interest in the image as soon as possible after it became present. A monitoring scheme with an associated stopping time will be proposed. The statistical goal will be to approximate the expected time to a false alarm of the scheme.

The new problem combines characteristics of a scanning statistic, in line with the second basic example that was used in the first part of the book, with characteristics of sequential monitoring for a change. The basic element that will be examined in order to carry out the analysis will be a random field. This random field will reflect both the scanning statistic component of the problem and its time-sequential component. The resulting field will be, accordingly, more complex. Furthermore, in order to add interest to the problem we formulate it in a way that makes the local statistical characteristics of the random field in the directions associated with the scanning statistic element different than the local characteristics in the direction associated with sequential monitoring.

To add insult to injury we use, as the basic monitoring scheme, an alternative to the cusum procedure. This alternative scheme is called the Shiryaev–Roberts procedure. In the context of the simple change-point detection problems that were discussed in Chapter 4, the Shiryaev–Roberts monitoring index emerges as the sum of likelihood ratios (exponentiated log-likelihoods) instead of the maxima. Hence, if $q_n = \max_{1 \le k \le n} \ell_{k,n}$ is the cusum monitoring index then the parallel Shiryaev–Roberts index is $r_n = \log \sum_{k=1}^{n} e^{\ell_{k,n}}$, with a stopping time that declares a change once the process r_n crosses a threshold. The motivation behind this alternative approach is Bayesian statistical philosophy that treats parameters as random variables. It can be shown that under an appropriate formulation of optimality the Shiryaev–Roberts procedure is asymptotically optimal. Energized by this optimality statement, that is claimed but will not be proved here, we will consider the properties of statistics that are produced by the summation of the exponentiated field with respect to the change-point parameter k. Likewise, the parameters associated with the direction of scanning will be integrated with respect to a density that reflects our prior beliefs regarding the characteristics of the signal. Maximization still takes place with respect to the date of monitoring n, considered as a parameter.

The last deviation in this application from the problems we considered before is with respect to the model of random noise. Previously we used the Gaussian model of white noise in the context of scanning statistic. That produced a Gaussian field that was used in the first part of the book. Currently, we will be using a sequence of independent two-dimensional Poisson random fields with a fixed baseline rate for each. Each position in the field represents a pixel. Each pixel can be randomly activated, with the number of active pixels in a given region following the Poisson distribution with a rate that equals λ times the area of the region. The distributions in disjoint regions are independent.

A signal is expressed in the form of a structured elevation in the rate. The parameters that specify the signal are the location, the width, and the intensity of the elevation. In order to simplify the notations and reduce the number of parameters involved we will use a rotationally symmetric signal so the issue of the orientation of the signal is eliminated.

It may be remarked that even if we had decided to maintain the Gaussian formulation we would not have been in a much better shape in terms of the ability to use classical tools for Gaussian fields. These tools in their standard application are designed to deal with maxima in fields, not with summation or integration of exponentiated fields as we have in our current application.

In the next section we describe in detail the statistical model that we propose to the problem and the monitoring scheme for the quickest detection of the emergence of a signal in the image. The section that follows is devoted to the analysis of the statistical properties of the scheme. In the last section we discuss very briefly the issue of optimal detection and methods for investigating optimality and approaches for constructing optimal procedure.

8.2   The statistical model

The movie frames, sometimes referred to as the monitoring periods, are denoted by i, $i = 1, 2, \ldots$. An image is composed of a continuum collection of pixels $\mathcal{P}$. At each monitoring period i an entire Poisson random field X_i is recorded. We assume that the baseline rate for all the fields is known and constant and denote it by λ.

The presence of a signal of a given shape and intensity in an image is reflected in the rate of the pixel measurements of which the image is composed. Suppose that the source of the signal is at position t. For a measurement taken at that position the intensity is $\lambda e^{\beta} dp$, for some parameter $\beta > 0$ that characterizes the relative intensity of the signal. The intensity at measurements near the center of the signal is relatively high. This intensity is reduced as we move away from the center. For pixels that are far enough from the center the intensity will practically be indistinguishable from the baseline intensity. We may model this description with the aid of a kernel function that equals 1 at the origin and decreases to 0 away from the origin. An example of such a function is $g(y) = e^{-\frac{1}{2}\|y\|^2}$, where the norm is the standard Euclidean norm in $\mathbb{R}^2$. Consequently, the model may propose an alternative rate at pixel j given by $\lambda_t(dp) = \lambda \exp\{\beta g((p - t)/h)\}dp$, for some $h > 0$ that represents the width of the domain of influence of the source.

For illustration consider the images shown in Figure 8.1. They are associated with the example given above. The grid has $2^5 \times 2^5 = 1024$ pixels in the unit square, $t = (0.8, 0.5)$, $h = 1/\sqrt{50}$, and $\beta = 2^{1/2}$. In Figure 8.1(a) the signal is plotted. It should be noted that the height of the peak of the signal is approximately 0.253. A typical image is obtained by generating 1024 independent Poisson random variables with a rate indicated by the given parameters. The result is shown in Figure 8.1(b). There is a 75% chance of obtaining observations larger than 3. Not surprisingly, the presence of a signal is difficult to detect. Figure 8.1(c) is obtained by averaging 20 independent images that contain the signal. Knowing the presence and location of the signal, we observe an elevated region associated with the signal. The presence and the location of the signal become more apparent after the averaging of 100 independent images. This is shown in Figure 8.1(d).

The goal of sequential change-point detection is to detect the presence of a signal as soon as possible after its emergence. No prior knowledge is assumed about the time of emergence or of the location of the signal. All procedures that are mentioned in this chapter detect the given signal with high probability well before the 20th observation. Yet some procedures may perform better than others. In particular, the procedure that we put our focus on is relatively advantageous.

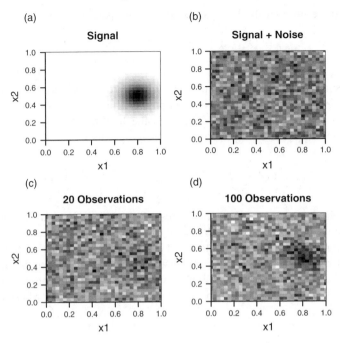

Figure 8.1   An example. Two-dimensional grid of observations is $2^5 \times 2^5$, $t = (0.8, 0.5)$, $h = 1/\sqrt{50}$, $\beta = 2^{1/2}$. (a) gives just the signal. In (b-d) independent Poisson noise with unit standard deviation have been generated.

The model will become more complex momentarily. Thereby, in order to make the presentation more manageable we will consider the parameters β and h to be fixed constants and treat only the two-dimensional t as a true parameter. Moreover, in order to avoid the need to deal with edge effects we will assume that the Poisson random field extends indefinitely.

Consider inner products. The inner product $\langle X, g \rangle$ corresponds to the random variable $\int g(p) X(dp)$ and the inner product $\langle f, g \rangle$ corresponds to the number $\int g(p) f(p) dp$.

One can associate the signal with a function $g_t : \mathcal{P} \to \mathbb{R}$, given by $g_t(p) = \beta g((p - t)/h)$, $p \in \mathcal{P}$. Use the fact that the log-likelihood ratio of a Poisson observation $X_i(dp)$ is $X_i(dp) \cdot \log(\lambda e^{g_t(p)}/\lambda) - \lambda e^{g_t(p)} dp + \lambda dp$ and the fact that the log-likelihood ratio of independent increments is the integral of the individual components to conclude that the log-likelihood ratio statistic at period i for testing the null hypothesis against the alternative is:

$$\langle X_i, g_t \rangle - \lambda \langle \exp g_t - 1, 1 \rangle ,$$

with 1 the constant function that assigns the value 1 across $\mathcal{P}$. For a change that occurs at period k and is assessed based on the data that has accumulated by the

nth period we get that the log-likelihood ratio statistic is:

$$\ell_{k,n,t} = \langle X_{k,n}, g_t \rangle - (n - k + 1)\lambda \langle \exp g_t - 1, 1 \rangle ,$$

for $X_{k,n} = \sum_{i=k}^{n} X_i$.

The Shiryaev–Roberts index that we would like to consider for this problem selects a prior distribution with compact support for the location of the signal t, say in the form of a density ρ, integrates the likelihoods with respect to that distribution and sums over the potential change-points k, $1 \leq k \leq n$:

$$r_n = \log \sum_{k=1}^{n} \int e^{\langle X_{k,n}, g_t \rangle - (n-k+1)\lambda \langle \exp g_t - 1, 1 \rangle} \rho(t) dt .$$

The associated stopping time sets a threshold x and declares a discovery once the threshold has been crossed: $N_x = \inf \{n : r_n \geq x\}$.

In the next section we investigate the statistical properties of the stopping time under the assumption that all observations emerge from the null Poisson distribution with a rate λ. Initially, we analyze the probability of stopping within a short interval and then we extend to the entire distribution and its expectation.

8.3   Analysis of statistical properties

In this application we are blessed with a confusing abundance of likelihood ratios to choose from. Not only do we have the parameter-specific log-likelihoods $\ell_{k,n,t} = \langle X_{k,n}, g_t \rangle - (n - k + 1)\lambda \langle \exp g_t - 1, 1 \rangle$ that we may use but also:

$$e^{\ell_{k,n}} = \int e^{\ell_{k,n,t}} \rho(t) dt$$

is a likelihood ratio. (This follows from the fact that it is non-negative and has an expectation of 1 under the null distribution P.) In fact, it is the likelihood ratio that involves the marginal distribution of the data stored in the frames $X_k, \ldots, X_n$. The marginal is with respect to the joint distribution in the Bayesian model that assigns the distribution ρ to the parameter t. By extension, this alternative distribution can be considered as a distribution of the entire sequence of frames. This extended distribution assigns the null distribution to the frames before period k and after period n. The frames between k and n are assigned the marginal Bayesian distribution. Naturally, we call this alternative distribution $P_{k,n}$. This alternative distribution should be distinguished from the distribution $P_{k,n,t}$ which is associated with $\ell_{k,n,t}$. One may note that while the frames are independent under the distribution $P_{k,n,t}$ they are not so, at least with respect to the frames between k and n, under the distribution $P_{k,n}$.

8.3.1   Preliminary localization

Let us use these alternative distributions and likelihood ratios in order to identify the factor associated with the large deviation rate. Start by considering marginal probabilities of the form $P(\ell_{k,n,t} \geq x)$. This setting is practically the same setting that was discussed in Section 3.2.1 in the context of the sequential probability test and later in Section 4.3 in the context of the cusum procedure. We found that this marginal probability is maximized when the expectation $E_{k,n,t}(\ell_{k,n,t})$ is equated with x. The rate of convergence to zero of the maximal probability is e^{-x} (divided by the square root of the number of observations and multiplied by a constant). The marginal probability remains significant as long as the difference between the expectation and the threshold is $O(\sqrt{x})$.

The expectation in the current setting is:

$$E_{k,n,t}(\ell_{k,n,t}) = \langle E_{k,n,t}(X_{k,n}), g_t \rangle - (n - k + 1)\lambda\langle e^{g_t} - 1, 1 \rangle$$
$$= (n - k + 1) \times (\lambda[\langle e^{g_t}, g_t \rangle - \langle e^{g_t} - 1, 1 \rangle]) = (n - k + 1) \times I .$$

Observe that with the assumptions we made the Kullback–Leibler information index I does not depend on t.

We can extend the result to the marginal probabilities associated with the other log-likelihood ratios $\ell_{k,n}$. Heuristically, we use the same approach based on the likelihood ratio:

$$P(\ell_{k,n} \geq x) = e^{-x}E_{k,n}(e^{-[\ell_{k,n}-x]}; \ell_{k,n} \geq x) .$$

However, this time we cannot just quote a standard local limit theorem in order to justify the approximation of the expectation since the statistic is not a sum of independent random variables under the alternative distribution. Yet, when we switch back to a representation that involves the distribution $P_{k,n,t}$ then we may show that the statistic is approximately such a sum.

Use the Bayesian interpretation of the distribution $P_{k,n}$ and condition on the distribution of the parameter t:

$$e^{-x}E_{k,n}(e^{-[\ell_{k,n}-x]}; \ell_{k,n} \geq x) = e^{-x} \int E_{k,n,t}(e^{-[\ell_{k,n}-x]}; \ell_{k,n} \geq x)\rho(t)dt .$$

Analyze each of the expectations inside the integral. They involve the random variable $\ell_{k,n}$. This random variable is produced by integration but it is dominated by the parameter-specific log-likelihood:

$$\ell_{k,n} = \log \int e^{\ell_{k,n,s}}\rho(s)ds = \ell_{k,n,t} + (\ell_{k,n,\hat{t}} - \ell_{k,n,t}) + \log \int e^{\ell_{k,n,s} - \ell_{k,n,\hat{t}}}\rho(s)ds ,$$

where $\hat{t} = \arg.\max \ell_{k,n,s}$ is the maximum likelihood estimator of the parameter t on the basis of the data stored in the $n - k + 1$ frames. The parameter specific log-likelihood $\ell_{k,n,t}$ is a good approximation of the mixture log-likelihood $\ell_{k,n}$ if the two remaining terms are relatively small.

For large values of the threshold x we will get that the typical number of frames that are involved in the computation of the log-likelihood ratios, i.e., $n - k + 1$, is also large. Standard large sample theory can teach us what to expect of the stochastic behavior of the log-likelihoods and the maximum likelihood estimator $\hat{t}$. As a corollary of this theory, together with a Laplace approximation of the integral in the last term, we will obtain an assessment of the magnitude of the discrepancy between $\ell_{k,n}$ and $\ell_{k,n,t}$.

The gradient of the log-likelihood $\dot{\ell}_{k,n,t}$ is denoted the score function and plays a key role. Under the $P_{k,n,t}$ distribution, this random vector has a zero mean and variance-covariance matrix which is equal to the Fisher information $-E_{k,n,t}[\ddot{\ell}_{k,n,t}] = (n - k + 1)\Psi$. Carry out a first-order Taylor expansion of the score function about $\hat{t}$ to obtain that:

$$\dot{\ell}_{k,n,t} \approx [-\ddot{\ell}_{k,n,\hat{t}}](\hat{t} - t) \ .$$

Notice that the score evaluated at $\hat{t}$ is equal to the zero vector. As a result, we get an approximate representation of the deviation between the maximum likelihood estimator and the parameter value it seeks to estimate. This representation is given in terms of the score function:

$$\hat{t} - t \approx [-\ddot{\ell}_{k,n,\hat{t}}]^{-1}\dot{\ell}_{k,n,t} \approx [(n + k - 1)\Psi]^{-1}\dot{\ell}_{k,n,t} \ .$$

The score has a zero expectation and a standard deviation proportional to the square root of the sample size. On the other hand, the Hessian of the log-likelihood is of the order of magnitude of the sample size. Therefore, when we divide one by the other we obtain that $\hat{t}$ converges to t. If so, due to continuity, we can replace $\hat{t}$ by t on the right-hand side of the approximation without changing the result by much. Add to that the fact that by the law of large numbers the average of the Hessian converges to its expectation to obtain the last approximation. That approximation tells us, as we apply the central limit theorem to the score function, that the difference $\hat{t} - t$, multiplied by the square root of the sample size, converges to the normal distribution with zero mean and variance-covariance matrix given by Ψ^{-1}.

If we go back and consider a two-term Taylor expansion of $\ell_{k,n,t}$ about $\hat{t}$, using the fact that the gradient vanishes at $\hat{t}$, we will now obtain that:

$$\ell_{k,n,\hat{t}} - \ell_{k,n,t} \approx (1/2)(t - \hat{t})'[\ddot{\ell}_{k,n,\hat{t}}](t - \hat{t}) \approx (1/2)[n^{\frac{1}{2}}(t - \hat{t})]'[\Psi][n^{\frac{1}{2}}(t - \hat{t})] \ .$$

The last random variable is asymptotically one-half of a chi-square random variable on 2 degrees of freedom. The bottom line is that the first of the two remaining terms is stochastically bounded hence negligible relative to $\ell_{k,n,t}$ that grows linearly with the sample size.

The approximation of the second remaining term follows from the application of the Laplace approximation of an integral. The Laplace approximation involves the maximal value with respect to s of the integrand $\exp\{\ell_{k,n,s}\}$, which is already removed in the current representation, and the curvature of the exponent part of the integrand in the vicinity of the maximizer. A second-order Taylor expansion, this time of $\ell_{k,n,s}$, will include this curvature as the leading term. The function ρ, if it is bounded from above in the entire range and bounded away from zero in the vicinity of the maximizer, can be evaluated at the maximizer. The resulting approximation is:

$$\int e^{\ell_{k,n,s}-\ell_{k,n,\hat{t}}}\rho(s)ds \approx \rho(\hat{t})\int e^{-\frac{n-k+1}{2}(s-\hat{t})'\left[\frac{\ddot{\ell}_{k,n,\hat{t}}}{n-k+1}\right](s-\hat{t})}ds \approx \frac{\rho(t)2\pi}{(n-k+1)\sqrt{|\Psi|}}.$$

Taking the log will identify that the second remaining term is asymptotic to $-\log(n-k+1)$, which is sub-linear.

The summary of this long and heuristic discussion is that under the $P_{k,n,t}$ distribution the marginal log-likelihood $\ell_{k,n}$ may be approximated by $\ell_{k,n,t}$ up to a deterministic term that is equal to the log of the number of observations and a stochastically bounded term. As a corollary we will obtain an asymptotic expansion of the marginal probability associated with $\ell_{k,n}$ which is similar to the expansion that was obtained for $\ell_{k,n,t}$. This heuristic statement relies on the fact that the Kullback–Leibler index and the Fisher information do not depend on t. In a case where they do, for example if h is treated as a parameter and assigned a prior distribution, then the asymptotic expansion for the two cases will not be the same. Typically, the exponential rate will not vary but the smaller order terms will.

The analysis given above can be made rigorous provided that the parameter-specific log-likelihood is smooth as a function of t. This is the case, for example, if the function that is used as a kernel is smooth. We selected a smooth kernel. If a different kernel is used then the conclusions should be reexamined.

After realizing that the large deviation factor is e^{-x} and that the range of interest involves likelihood ratios $\ell_{k,n}$ with $n-k+1 \in x/I \pm C\sqrt{x}$, for some large C, we can start the real work. At the heart of this work is the derivation of an approximation for a probability of the type $P(N_x \leq m)$, for $m = m(x)$ sub-exponential.

The stopping time N_x monitors the sequence $r_n = \log\sum_{k=1}^{n}e^{\ell_{k,n}}$. In order to restrict the parameter space to the region where action occurs we may want to redefine the monitoring sequence and the associated stopping time. A natural candidate is $\hat{r}_n = \log\sum_{k\in T_n}e^{\ell_{k,n}}$, $T_n = \{k : n-k+1 \in x/I \pm C\sqrt{x}\}$, where the summation is restricted to the range of parameters that matter. The stopping time that comes along with this sequence is $\hat{N}_x = \inf\{n : \hat{r}_n \geq x\}$. Our first task is to show that the chance of both stopping times to stop within a given interval in time is about the same.

We bound the probability of stopping with N_x, both from above and from below, by the probability of stopping with $\hat{N}_x$. One direction is easy. We always have that $\hat{r}_n \leq r_n$. Therefore, $N_x \leq \hat{N}_x$ and $P(\hat{N}_x \leq m) \leq P(N_x \leq m)$. So we want to bound the probability from above. Motivated by the analysis from

Chapter 4 we make the assumption that the probability associated with $\hat{N}_x$ is asymptotic to me^{-x}. If so, the examination of $\hat{N}_{x-\varepsilon}$ for $\varepsilon > 0$ small and fixed should give about the same answer. We will use the assumption, making a mental note to revise the proof that follows if the assumption turns out to be wrong.

Partitioning the sample space based on $\hat{N}_{x-\varepsilon}$ stopping in the given interval or not we have that:

$$P(N_x \leq m) \leq P(\hat{N}_{x-\varepsilon} \leq m) + P(\{N_x \leq m\} \cap \{\hat{N}_{x-\varepsilon} > N_x\}) .$$

Our goal is to show that the last probability is $o(me^{-x})$.

Let $e^{r_n} - e^{\hat{r}_n} = \sum_{k \in \bar{T}_n} e^{\ell_{k,n}}$, $\bar{T}_n = \{k \leq n : |n - k + 1 - x/I| > C\sqrt{x}\}$. We produce a bound on the probability via a sequential likelihood ratio identity:

$$P(\{N_x \leq m\} \cap \{\hat{N}_{x-\varepsilon} > N_x\}) = \sum_{n=1}^{m} P(\{N_x = n\} \cap \{\hat{N}_{x-\varepsilon} > n\})$$

$$= \sum_{n=1}^{m} E\left[\frac{\sum_{k \in \bar{T}_n} e^{\ell_{k,n}}}{\sum_{h \in \bar{T}_n} e^{\ell_{h,n}}}; N_x = n, \hat{N}_{x-\varepsilon} > n \right]$$

$$= \sum_{n=1}^{m} \sum_{k \in \bar{T}_n} E_k\left[\frac{1}{e^{r_n} - e^{\hat{r}_n}}; N_x = n, \hat{N}_{x-\varepsilon} > n \right],$$

where the expectation E_k is computed under the regime where a signal becomes present at period k according to the distribution ρ and stays thereafter.

Over the event $\{N_x = n\} \cap \{\hat{N}_{x-\varepsilon} > n\}$ we have that $r_n \geq x$ but $\hat{r}_n < x - \varepsilon$. Consequently, the difference in the denominator is no less than $e^x(1 - e^{-\varepsilon})$. As a result, the probability is bounded by:

$$\frac{e^{-x}}{1 - e^{-\varepsilon}} \sum_{n=1}^{m} \sum_{k \in \bar{T}_n} P_k(\{N_x = n\} \cap \{\hat{N}_{x-\varepsilon} > n\}) \leq \frac{e^{-x}}{1 - e^{-\varepsilon}} \sum_{n=1}^{m} \sum_{k \in \bar{T}_n} P_k(N_x = n) .$$

We change the order of summation and alow the summation in n to extend to infinity. This produces an upper bound on the probability of an error when we reduce the parameter set:

$$\frac{e^{-x}}{1 - e^{-\varepsilon}} \sum_{k=1}^{m} \{P_k(k \leq N_x < k + x/I - C\sqrt{x}) + P_k(N_x > k + x/I + C\sqrt{x})\} .$$

$$(8.1)$$

Bounding the probability will be completed once we show that the probabilities in the curely brackets can be made as small as we wish by selecting C to be large and fixed and by increasing x.

Start with the second probability in the curely brackets. For this probability we can use the fact that the monitoring sequence r_n is larger than each of the log-likelihood ratios that are used to form it. Consequently,

$$P_k(N_x > k + x/I + C\sqrt{x}) \le P_k(\ell_{k,k+x/I+C\sqrt{x}} < x)$$

$$= \int P_{k,t}(\ell_{k,k+x/I+C\sqrt{x}} < x)\rho(t)dt \ .$$

We evoke the approximation

$$\ell_{k,n} \approx \ell_{k,n,t} + U_2/2 + \log\{\rho(t)2\pi|\Psi|^{-\frac{1}{2}}\} - \log(n - k + 1)$$

and apply Chebishev inequality to the approximation in order to finish the job.

Bounding the first probability in the curly brackets of (8.1) requires more sophistication. Let

$$\sum_{h=1}^{n} e^{\ell_{h,n} - \ell_{k,n}} = \sum_{h=1}^{k} e^{\ell_{h,n} - \ell_{k,n}} + \sum_{h=k+1}^{n} e^{\ell_{h,n} - \ell_{k,n}} = W_{k,n} + R_{k,n} \ .$$

We consider the conditional distribution of the process, given $\mathcal{F}_{k-1}$, the σ-algebra generated by the first $k - 1$ frames. Notice that the event $\{N_x \ge k\} = \{N_x < k\}^c$ belongs to that σ-algebra. Clearly,

$$P_k(k \le N_x < k + x/I - C\sqrt{x}) \le P_k\left(\max_{0 \le n-k \le \frac{x}{I} - C\sqrt{x}} \ell_{k,n} \ge x - IC\sqrt{x}/2\right)$$

$$+ E_k\left[P_k\left(\max_{0 \le n-k \le \frac{x}{I} - C\sqrt{x}} W_{k,n} \ge e^{\frac{IC}{4}\sqrt{x}}\Big|\mathcal{F}_{k-1}\right); N_x \ge k\right]$$

$$+ P_k\left(\max_{0 \le n-k \le \frac{x}{I} - C\sqrt{x}} R_{k,n} \ge e^{\frac{IC}{4}\sqrt{x}}\right) \ .$$

We have three probabilities to consider. Start with the third one. Define $\nu_3 = \inf\{n : \log R_{k,n} \ge IC\sqrt{x}/4\}$ for the probability that involves $R_{k,n}$, $n \ge k$:

$$P_k(\nu_3 \le x/I - C\sqrt{x}) = \sum_{n=k}^{k+x/I-C\sqrt{x}} P_k(\nu_3 = n)$$

$$= \sum_{n=k}^{k+x/I-C\sqrt{x}} \sum_{h=k+1}^{n} E_h(1/R_{k,n}; \nu_3 = n)$$

$$\le e^{-IC\sqrt{x}/4} \sum_{h=k+1}^{k+x/I-C\sqrt{x}} \sum_{n=h}^{\infty} P_h(\nu_3 = n)$$

$$\le e^{-IC\sqrt{x}/4}(x/I - C\sqrt{x}) \ ,$$

which converges to 0 with the increase in x.

For the first probability we condition on the value of t and take:

$$\int P_{k,t} \left(\max_{0 \le n-k \le \frac{x}{I} - C\sqrt{x}} \ell_{k,n} \ge x - IC\sqrt{x}/2 \right) \rho(t)dt$$

$$\le \int P_{k,t} \left(\max_{0 \le n-k \le \frac{x}{I} - C\sqrt{x}} \ell_{k,n,t} \ge x - IC\sqrt{x}/4 \right) \rho(t)dt \qquad (8.2)$$

$$+ \int P_{k,t} \left(\max_{0 \le n-k \le \frac{x}{I} - C\sqrt{x}} (\ell_{k,n} - \ell_{k,n,t}) \ge IC\sqrt{x}/4 \right) \rho(t)dt \, . \qquad (8.3)$$

For the probability in (8.2), since $\ell_{k,n,t}$ is a random walk in n, one may use Kolmogorov's maximal inequality to give a uniform bound on the probabilities inside the integral. For the probabilities inside the integral in (8.3) one may define $\nu_{1,t} = \inf \{ n : \ell_{k,n} - \ell_{k,n,t} \ge IC\sqrt{x}/4 \}$ and apply a sequential likelihood ratio identity to obtain $\exp \{ -IC\sqrt{x}/4 \}$ as a uniform bound. The likelihood ratio identity is applicable since $\ell_{k,n} - \ell_{k,n,t}$ are log-likelihood ratios with respect to the distribution $P_{k,t}$.

There is still the second probability. It should be pointed out that the log-likelihood ratios $\ell_{h,n} - \ell_{k,n}$, for $n \ge k > h$, are log-likelihoods also under the conditional distribution, given the σ-algebra $\mathcal{F}_{k-1}$. We use this fact to use once more a sequential likelihood ratio identity. This time we set $\nu_2 = \inf \{ n : \log W_{k,n} \ge IC\sqrt{x}/4 \}$:

$$P_k(\nu_2 \le x/I - C\sqrt{x} | \mathcal{F}_{k-1}) = \sum_{n=k}^{k+x/I-C\sqrt{x}} P_k(\nu_2 = n | \mathcal{F}_{k-1})$$

$$= \sum_{n=k}^{k+x/I-C\sqrt{x}} \sum_{h=1}^{k} E(e^{\ell_{h,n}} / W_{k,n}; \nu_2 = n | \mathcal{F}_{k-1})$$

$$\le e^{-IC\sqrt{x}/4} \sum_{h=1}^{k} E(e^{\ell_{h,\nu_2}}; \nu_2 < \infty | \mathcal{F}_{k-1})$$

$$= e^{-IC\sqrt{x}/4} \sum_{h=1}^{k} e^{\ell_{h,k-1}} \, ,$$

since the conditional probability of eventually stopping is 1. You may also note that $\ell_{k,k-1} = 0$.

After multiplication by the indicator of the event $\{ N_x \ge k \}$ and taking expectations with respect to the P_k distribution (which equals the null distribution over the event) we get:

$$P_k \left(\max_{0 \le n-k \le \frac{x}{I} - C\sqrt{x}} W_{k,n} \ge e^{\frac{IC}{4}\sqrt{x}} \right) \le e^{-IC\sqrt{x}/4} \sum_{h=1}^{k} P_h(N_x \ge k) \, .$$

One last push and we are there. Clearly, $\sum_{i=1}^{n} e^{\ell_{i,n}} \geq \sum_{i=h}^{n} e^{\ell_{i,n}}$. It follows that $P_h(N_x \geq k) \leq P_1(N_x \geq k - h + 1)$. Consequently,

$$\sum_{h=1}^{k} P_h(N_x \geq k) \leq \sum_{h=1}^{k} P_1(N_x \geq k - h + 1) = E_1 \min\{v_x, k\},$$

for $v_x = \inf\{n : \ell_{1,n} \geq x\}$ a mixture-type sequential probability ratio test. This expectation is asymptotic to x and after multiplication by $\exp\{-IC\sqrt{x}/4\}$ it converges to 0.

Now we have established the fact that by choosing C large, and for all big enough threshold x, we can make the sum of the probabilities in the curly brackets in (8.1) as small as we wish in comparison with me^{-x}. We worked hard but our effort was not wasted since we have actually killed two birds with one stone. The term in (8.1) is a bound in probability between the cumulative distributions of N_x and $\hat{N}_{x-\varepsilon}$, evaluated at m. The dependence on m in the bound is only through the range of summation and not through the probabilities in the curly brackets. The implication is that the bound is valid for the entire distribution of the stopping times, as long as m is proportional to e^x. We use this bound now in order to justify the restriction to a smaller subset of parameters when we analyze the probability of stopping in a small interval of time. We will be able to reuse the same bound later on in order to justify the similarity of the entire distribution functions for $\hat{N}_x$ and for N_x.

8.3.2 Measure transformation, the localization theorem, and integration

We are heading now to the production of the likelihood ratio identity representation of the probability as a sum of expectations. Denote the reduced parameter space by:

$$T = \{(k, n) : x/I - C\sqrt{x} \leq n - k + 1 \leq x/I + C\sqrt{x}\}$$

and define $T_n = \{k : (k, n) \in T\}$. This is the same parameter space and partition thereof that was used for the cusum in Chapter 4. Like then, we will use the sum of likelihoods $\sum_{(k,n)\in T} e^{\ell_{k,n}}$ for the likelihood identity. The difference is that the likelihoods are of a mixture-type. Still, they are valid likelihood ratios.

The likelihood ratio identity is:

$$P(\hat{N}_x \leq m) = e^{-x} \sum_{(k,n)\in T} E_{k,n}\left(\frac{M_{k,n}}{S_{k,n}} e^{-[\tilde{\ell}_{k,n}+m_{k,n}]}; \tilde{\ell}_{k,n} + m_{k,n} \geq 0\right),$$

where $S_{k,n} = \sum_{l=1}^{m} \sum_{h\in T_l} e^{\ell_{h,l}-\ell_{k,n}}$ and $M_{k,n} = \max_{1\leq l\leq m} \sum_{h\in T_l} e^{\ell_{h,l}-\ell_{k,n}}$ are the sum, and the new maximal term, of likelihood ratios. Also, $m_{k,n} = \log M_{k,n}$, and $\tilde{\ell}_{k,n} = \ell_{k,n} - x$. Note the slight variation in the definition of $M_{k,n}$ to accommodate the fact that the monitoring statistic is constructed by summing likelihood ratios.

Another difference between the analysis that was produced for the cusum and the current analysis is that we add a step of conditioning on the value of t. However, some probabilities are easier to handle in the $P_{k,n}$ distribution. Anticipating future difficulties, we would like to truncate the random variable $m_{k,n}$ before we condition on the value of t. Specifically, for the event $\{m_{k,n} \geq c \log x\}$ we have that:

$$P_{k,n}(m_{k,n} \geq c \log x) \leq P_{k,n}(S_{k,n} \geq x^c) \leq |T| x^{-c}.$$

The number of parameters in T is proportional to $m\sqrt{x}$. If m is polynomial in x we may take c to be 2 plus the degree of the polynomial in order to eliminate any worries. Henceforth, we will assume that that $m_{k,n} \leq c \log x$.

Let us do the conditioning now:

$$E_{k,n}\left(\frac{M_{k,n}}{S_{k,n}} e^{-[\tilde{\ell}_{k,n} + m_{k,n}]}; \tilde{\ell}_{k,n} + m_{k,n} \geq 0\right)$$

$$= \int E_{k,n,t}\left(\frac{M_{k,n}}{S_{k,n}} e^{-[\tilde{\ell}_{k,n} + m_{k,n}]}; \tilde{\ell}_{k,n} + m_{k,n} \geq 0\right) \rho(t)dt .$$

The localization theorem will be applied to each of the expectations in the integral. Taking care of the integration will be part of the integration step at the end.

We saw that the leading term in $\ell_{n,k}$ is $\ell_{n,k,t}$, which has variance equal to $(n - k + 1)\lambda\langle e^{gt}, g_t^2\rangle = (n - k + 1)\sigma^2$. Consequently, it is natural to set $\kappa = n - k + 1$. We may also reuse the definition of the local σ-algebra $\hat{\mathcal{F}}_{n,k} = \sigma\{X_{k-\tau}, \ldots, X_{k+\tau}, X_{n-\tau}, \ldots, X_{n+\tau}\}$, this time interpreting the components X_i as containing the information of an entire frame.

We produce $\hat{M}_{k,n,t}$ and $\hat{S}_{k,n,t}$ by restricting the summation and maximization only to the range of parameter values that are at most τ away from either k or n. This is true also for the summation that appears as part of the definition of $M_{k,n}$. We also replace $\ell_{h,l} - \ell_{k,n}$ by their approximations $\ell_{h,l,t} - \ell_{k,n,t}$. As a result, the random variables $\hat{M}_{k,n}$ and $\hat{S}_{k,n}$ are measurable with respect to $\hat{\mathcal{F}}_{n,k}$ as required.

We still need to validate the conditions of Theorem 5.2. But before doing so we may want to entertain ourselves with the identification of the approximations that it will produce.

The term $\hat{M}_{k,n,t}$ can be represented as a product of a sum in the vicinity of k times a maxima in the vicinity of n. On the other hand, the term $\hat{S}_{k,n,t}$ is a product of two sums, one for k and one for n. When we take the ratio the sum associated with k cancels out and we are left with a single ratio. Taking expectations and letting $\tau \to \infty$ gives the term $E(\mathcal{M}/\mathcal{S})$. Interestingly enough, this term is the same term that emerges in the one-sided sequential probability ratio test that is discussed in Chapter 2. Specifically, the increments in the sequential test emerge from testing the null hypothesis of a constant rate λ over a frame against the alternative that the Poisson rate is λe^{gt}, for the given t. By construction we obtain that the constant in the current case does not depend on the selection of t. It is not the same constant that emerges in the case of testing for a shift in a Gaussian mean. Yet, using the simple ad hoc formula (2.4) for the Gaussian case, instead of the actual limit that is given in (2.3) and refers to the distribution of the partial sums produced by a weighted combination of Poisson random variables, will probably not cause an enormous numerical error.

Overall we may expect the conclusion of the localization theorem to produce:

$$P(\hat{N}_x \leq m) \approx e^{-x} \sum_{(k,n) \in T}$$

$$\times \int \frac{1}{\sqrt{(n-k+1)\sigma^2}} \phi\left(\frac{I(n-k+1)-x}{\sqrt{(n-k+1)\sigma^2}}\right) E[\mathcal{M}/\mathcal{S}]\rho(t)dt$$

and upon integration to lead to the limit:

$$\lim_{x \to \infty} (e^x/m)P(\hat{N}_x \geq x) = \frac{1}{I}E[\mathcal{M}/\mathcal{S}][2\Phi(CI^{\frac{3}{2}}\sigma^{-1}) - 1].$$

The simple form of the limit is partially a result of the fact that the terms $E[\mathcal{M}/\mathcal{S}]$, I and σ^2 are independent of t. If they would have been dependent on the Bayesian parameter then the right-hand side of the equation should have been integrated with respect to the density ρ in order to produce the constant term.

Obviously, the parallel constant for the unrestricted N_x will be $E[\mathcal{M}/\mathcal{S}]/I$, or the integral thereof if this term depends on the Bayesian parameter.

We can even go further than that. The Poisson argument can be lifted as is from the discussion of the cusum procedure in Chapter 4 to extend the approximation of the distribution of $\hat{N}_k$ to the entire range and obtain an exponential limit. In light of the approximation proposed by (8.1), we can establish an exponential limit distribution for the unrestricted stopping time N_x. Based on dominated convergence theorem, using again the same argument as in the case of the cusum, we will get an approximation for the average run length to false alarm for the Shiryaev–Roberts change-point detection procedure:

$$\lim_{x \to \infty} e^{-x}E(N_x) = I/E[\mathcal{M}/\mathcal{S}]. \tag{8.4}$$

8.3.3   Checking the conditions of the localization theorem

It is time to check conditions for the localization theorem before we celebrate the accomplishment of this chapter's main task. Condition I* is satisfied with $C = 1$. Condition III* is obtained from the fact that $E[\hat{M}_\kappa/\hat{S}_\kappa] = E[\hat{\mathcal{M}}/\hat{\mathcal{S}}]$ and the convergence of $E[\hat{\mathcal{M}}/\hat{\mathcal{S}}]$ with the increase in τ.

8.3.4   Checking Condition V*

For Condition V* one can ignore the absolute value sign since $M_{k,n}$ and $\hat{M}_{k,n}$ are larger than 1. The condition for the first probability is met with $c \log x$ since we bounded that random variable before. Unlike the examples that we considered up until now, it is not sufficient to check only the first probability since there is no monotone relation between $M_{k,n}$ and $\hat{M}_{k,n}$. Still, checking the condition for the other two probabilities is not hard.

For the second probability we have that $P_{k,n,t}(\log \hat{M}_{k,n,t} \geq c \log x) \leq 2\tau x^{-c}$. For the third probability we do have monotonicity between $M_{k,n}$ and a parallel term that uses the mixture log-likelihood ratios $\ell_{h,l} - \ell_{k,n}$, but with h restricted to be within τ of k and l to be within τ of n. The third probability will follow from the comparison of this reduced term and $\hat{M}_{k,n}$. This comparison, in turn, follows from the approximation of the mixture-type log-likelihood ratios by the parameter specific log-likelihood ratios. As said, this approximation is produced by large sample consideration and Laplace approximation:

$$\ell_{h,l} - \ell_{k,n} \approx \ell_{h,l,t} - \ell_{k,n,t}$$
$$+ (1/2)(t - \hat{t}_{h,l})'[\ddot{\ell}_{h,l,\hat{t}_{h,l}}](t - \hat{t}_{h,l}) - (1/2)(t - \hat{t}_{k,n})'[\ddot{\ell}_{k,n,\hat{t}_{k,n}}](t - \hat{t}_{k,n})$$
$$+ \log\{\rho(\hat{t}_{h,l})/\rho(\hat{t}_{k,n})\} - (1/2)\log\{|\ddot{\ell}_{h,l,\hat{t}_{h,l}}|/|\ddot{\ell}_{k,n,\hat{t}_{k,n}}|\}$$

The maximum likelihood estimators $\hat{t}_{h,l}$ and $\hat{t}_{k,n}$ are constructed using, practically, the same collection of frames. Thereby, they are almost equal to each other. The Hessian matrices $\ell_{h,l,\hat{t}_{h,l}}$ and $\ell_{k,n,\hat{t}_{k,n}}$ are proportional to $l - h + 1$ and $n - k + 1$, respectively, and they are continuous as a function of the parameter t. Consequently, the the ratio of the determinants converges almost surely to 1. It follows that the error in the approximation is more than some given quantity only if the distance between $\hat{t}_{h,l}$ and $\hat{t}_{k,n}$ is more than another small, but fixed, number. The probability of such a discrepancy between the estimators is exponentially small as a function of the sample size, small enough to satisfy the condition.

8.3.5   Checking Condition IV*

For the proof of Condition IV* we use once more the approximation of $\ell_{k,n}$ by the sum of $\ell_{k,n,t}$, a random element which is a function of the standardized maximum likelihood estimator, the log of the number of frames that are used, and a term that depends on the maximum likelihood estimator and converges to a constant.

Specifically, we condition on the local σ-algebra $\mathcal{F}_{k,n}$ and consider the joint distribution of the independent sum $\ell_{k+\tau+1,n-\tau-1,t}$ and the maximum likelihood estimator $\hat{t}_{k,n}$. The standardized maximum likelihood estimator is asymptotically equivalent to $(n - k + 1)^{-\frac{1}{2}}\Psi^{-1}\ddot{\ell}_{k,n,t}$ and therefore, as we will see momentarily, asymptotically uncorrelated with $\ell_{k,n,t}$. Therefore, we are in a position to apply Theorem 5.3 with respect to this joint distribution. Accordingly, we obtain a local limit with respect to $\ell_{k+\tau+1,n-\tau-1,t}$ that is not affected by the perturbation cased by the random discrepancy of it from $\ell_{k,n}$, since this discrepancy is determined by $\hat{t}_{k,n}$.

In order to check that the parameter specific log-likelihood and the score are independent under the alternative observe that

$$(\partial/\partial t)\langle X_i, g_t\rangle = \langle X_i, \mathring{g}_t\rangle ,$$

where $\mathring{g}_t : \mathcal{P} \to \mathbb{R}^2$ is the function that assigns to the point p the value $(g_t(p)(p_1 - t_1), g_t(p)(p_2 - t_2))'$. The covariance between the gradient vector and $\langle X_i, g_t\rangle$ is, as required, equal to:

$$\mathrm{Cov}_t(\langle X_i, \mathring{g}_t\rangle, \langle X_i, g_t\rangle) = \langle \lambda g_t, g_t \mathring{g}_t\rangle = \lambda \int g_t^3(p)(p - t)dp = 0 .$$

8.3.6 Checking Condition II*

Last we come to Condition II*. This condition may be investigated by the consideration of events such as $\{\ell_{h,l} - \ell_{k,n} \ge \log(\varepsilon p_{h,l})\}$, intersected or not with the event $\{\tilde{\ell}_{k,n} + y \in [-m, \delta]\}$, for some finite m and δ. Recall that $\ell_{k,n}$ can be approximated by $\ell_{k,n,t}$ up to a deterministic term of the order $\log(n - k + 1)$ and a random term of the order $O_p(1)$. The main issue is how to deal with $\ell_{h,l}$ in the environment determined by the parameters k, n, and t.

When all but a small proportion of the frames that produce the two log-likelihoods are shared then one may use the upper bound:

$$\ell_{h,l} - \ell_{k,n} \le \ell_{h,l,\hat{t}_{h,l}} - \ell_{k,n,\hat{t}_{k,n}} .$$

With a very high probability the two estimators $\hat{t}_{h,l}$ and $\hat{t}_{k,n}$ differ from each other by no more than a small amount. Since the $\hat{t}_{k,n}$ is close to t this implies that so is $\hat{t}_{h,l}$. Consequently, with sufficiently high probability, the analysis essentially reduces to dealing with $\ell_{h,l,t} - \ell_{k,n,t}$. This case is practically identical to the analysis that involved the sequential probability ratio test in Chapter 3.

However, when the symmetric difference between the interval $[h, l]$ and $[k, n]$ contains more than a small proportion of the total then the analysis is more complex. This issue can be resolved if we can show that typically the difference $\ell_{h,l} - \ell_{k,n,t}$ is less than $(-\varepsilon)$ times the number of frames in the symmetric difference between $[h, l]$ and $[k, n]$, namely the number of frames that are involved in the computation in one statistic but not in the other. The probability of the event where this inequality regarding the difference $\ell_{h,l} - \ell_{k,n,t}$ does not hold should be exponentially small as a function of the number of frames in the symmetric difference between the intervals.

Start with the case where $[h, l]$ and $[k, n]$ are disjoint. In such a case we have that the log-likelihood $\ell_{h,l}$ is examined in a region where the null hypothesis rules:

$$P_{k,n,t}(\ell_{h,l} - \ell_{k,n,t} \ge -y) \le P(\ell_{h,l} \ge y) + P_{k,n,t}(-\ell_{k,n,t} \ge -2y)$$

$$\le e^{-y} + e^y E_{k,n,t}(e^{-\frac{1}{2}\ell_{k,n,t}})$$

$$\leq e^{-y} + e^{y}\mathrm{E}(e^{\frac{1}{2}\ell_{k,n,t}})$$

$$\leq e^{-y} + e^{y}e^{-\frac{1}{2}\lambda(n-k+1)\langle(\exp\{\frac{1}{2}g_t\}-1)^2,1\rangle} .$$

Thus we can select y to be proportional to $n - k + 1$ and obtain an exponential convergence to 0.

Next consider the case where the interval $[h, l]$ is contained inside the interval $[k, n]$. Using a similar argument:

$$\mathrm{P}_{k,n,t}(\ell_{h,l} - \ell_{k,n,t} \geq -y) \leq \mathrm{P}(\ell_{h,l} - \ell_{h,l,t} \geq y) + \mathrm{P}_{k,n,t}(\ell_{h,l,t} - \ell_{k,n,t} \geq -2y)$$

$$\leq e^{-y} + e^{y}e^{-\frac{1}{2}\lambda(h-k+n-l)\langle(\exp\{\frac{1}{2}g_t\}-1)^2,1\rangle} .$$

This time one can select y to be proportional to $h - k + n - l$, the number of frames that belong to $[k, n]$ but not to $[h, l]$.

A slightly more involved argument can be used for the case where $[h, l]$ contains the interval $[k, n]$. Set $l_{h,l} = \log \int e^{\ell_{h,l,s}-\ell_{k,n,s}} \rho(s)ds$. The ratio:

$$e^{\ell_{h,l}-l_{h,l}} = \int e^{\ell_{k,n,t}} \left[\frac{e^{\ell_{h,l,t}-\ell_{k,n,t}}\rho(t)}{\int e^{\ell_{h,l,s}-\ell_{k,n,s}}\rho(s)ds}\right] dt$$

is a likelihood ratio for the observations over the interval $[k, n]$. The null distribution of these observation is the original null distribution. The alternative distribution assigns a random prior to the location t and computes the marginal joint distribution of the frames in the given interval $[k, n]$ on the basis of this prior. The prior distribution is determined by the frames that belong to $[h, l]$ but not to $[k, n]$. These frames are independent of the frames in $[k, n]$ under the null distribution. We may use the fact that

$$\mathrm{P}_{k,n,t}(\ell_{h,l} - \ell_{k,n,t} - l_{h,l} \geq y) \leq e^{-y}\mathrm{E}_{k,n,t}\left[\frac{e^{-\ell_{k,n,t}}e^{\ell_{h,l}}}{e^{l_{h,l}}}\right] = e^{-y}\mathrm{E}\left[\frac{e^{\ell_{h,l}}}{e^{l_{h,l}}}\right] .$$

The last expectation is an expectation of a likelihood ratio and is equal to 1. It follows that:

$$\mathrm{P}_{k,n,t}(\ell_{h,l} - \ell_{k,n,t} \geq -y) \leq e^{-y} + \mathrm{P}_{k,n,t}(l_{h,l} > -2y) = \mathrm{P}(l_{h,l} > -2y) ,$$

since $l_{h,l}$ is a function of frames that do not belong to the interval $[k, n]$.

In order to bound the distribution of the mixture-type likelihood $l_{h,l}$ we cover the compact support of ρ with a finite collection of subsets T_j: $K \subset \cup_j T_j$. The subset T_j is centered at t_j and has a small radius. Denote by $\rho_j = \rho(T_j)$ the probability assigned by ρ to T_j. Now:

$$\mathrm{P}(l_{h,l} > -2y) \leq \sum_j \mathrm{P}\left(l_{h,l,t_j} + \max_{t \in T_j}(l_{h,l,t} - l_{h,l,t_j}) \geq -2y + \log\rho_j\right)$$

$$\leq \sum_j \mathrm{P}(l_{h,l,t_j} \geq -3y + \log\rho_j) + \sum_j \mathrm{P}\left(\max_{t \in T_j}\langle Y, g_t - g_{t_j}\rangle \geq y\right),$$

for $Y = \sum_{i \in [h,l] \setminus [k,n]} (X_i - \lambda \cdot 1)$. By setting the sets T_j to be small enough we may assure, by the properties of the normal kernel, that each of the maximal random variables has a moment generating function over an open set that contains the origin. This assures an exponentially decreasing tail and completes the proof for this case.

For the last case where the intervals $[k, n]$ and $[h, l]$ intersect but one is not included in the other we may apply one argument for the part of the symmetric difference that contains frames from $[k, n]$ and the other argument for the part of the symmetric difference that contains frames from $[h, l]$. At least one of the two parts should contain a large number of frames, enough to derive the exponential decay. This completes the verification of Condition II* and the proof of the localization theorem.

8.4   Optimal change-point detection

In the current chapter we investigated the average run length to false alarm of a particular change-point detection rule, the Shiryaev–Roberts procedure that uses Bayesian, or mixture log-likelihood ratio statistics. One can easily envision other candidates for monitoring quality indexes and stopping rules. For example, one can propose to estimate the change-point, i.e., maximize the log-likelihood instead of summing the exponentiated log-likelihoods. The result will be a cusum statistic $\max_k \ell_{k,n}$ that uses mixture log-likelihood ratio statistics. One can go even further than that and propose to estimate the location using a frequentist's maximum likelihood approach instead of a Bayesian mixture approach and obtain $\max_{k,t} \ell_{n,k,t}$ as the basic monitoring index. The average run length to false alarm for each of these procedures may be analyzed using random field considerations, since the probabilities involved emerge as extreme probabilities in appropriate random fields. Consequently, in principle we may apply the method that was developed in order to obtain the statistical properties of the different proposals.

The options are limitless. But is there an objective criterion that may help us choose among the options? We may relate this problem to a similar problem that emerges in the context of hypothesis testing. A statistical test is permissable if its significance level is no more than a pre-selected value of α. However, many tests obey that minimal criterion, some of which are obviously useless. Such a useless test is the admissible test that ignores the data and decides to reject or not on the basis of flipping a coin. (An appropriately biased coin to assure that rejecting the null hypothesis occurs with probability α.) So we may ask the same question again, formulated this time in the context of hypothesis testing: is there an objective criterion to choose among admissible tests?

Answers to the question in the context of hypothesis testing are typically formulated in terms of the statistical power. Statistical power is the probability to reject the null hypothesis – thus accept the alternative hypothesis – when the actual distribution of the data is indeed from the alternative hypothesis. For two admissible tests, the one with the greater statistical power is preferred. Of course, this general rule is only a guideline. If the better test is much harder to implement, say it involves difficult computations, and the loss in power due to the use of a much simpler procedure is small then one may prefer the simpler test after all.

A similar philosophy guides the process of selecting a change-point detection procedure. Average run length to false alarm plays the role that was occupied by the significance level in hypothesis testing. The substitute for the concept of statistical power is the concept of average run length to detection, known as detection delay. All competing procedures are required to obey the average run length to false detection, namely to have that the expectation of the stopping rule when no change occurs is set at a given level. Among all permissible procedures the ones which require, on the average, the accumulation of less post-change observations prior to setting an alarm are preferred. As in the case of hypothesis testing, the ordering proposed by this criterion may be modified by other considerations such as computational efficiency.

Considerations of statistical efficiency, the power in hypothesis testing and the average detection delay in sequential change-point detection, become more complicated when there is more than one alternative distribution to choose from. The actual distribution of the post change observations can be associated with any value for the parameters that were considered when modeling the alternative distribution. In fact, the distribution of these observations may not even be part of the model that was considered for the construction of the procedure.

This issue is even more complicated in change-point detection where the time of change – yet another parameter – is a priori unknown. Worse than that, there is no agreement in the literature on which form of expectation to use in the computation of the detection delay. A Bayesian approach examines the difference $N - k + 1$, where N is the stopping time and k is the time of change, as a difference between two random variables. Holding firm to the belief that all parameters are actually random variables, it maybe proposed to measure the expected delay in detection as an expectation over the joint distribution of k and N: $E(N - k + 1; N \geq k)$. The control on the rate of false alarm is obtained by placing a constraint on the probability $P(N < k)$ of stopping prior to the change. Indeed, a celebrated result by Shiryaev states that an appropriate version of the Shiryaev–Roberts procedure is optimal under this formulation if the prior distribution of the time of change is geometric [23].

If one has no prior assumption at what time a change is likely to occur then one may want to use a different criterion. The criterion of choice for many frequentists is to guard against the worst case scenario and use: $\max_k E_k(N - k + 1 | N \geq k)$. Building on the results of Shiryaev, Pollak was able to show that in the absence of any other parameters the Shiryaev–Roberts procedure is asymptotically optimal [24]. Of special interest in the context of this mini-max formulation are stopping rules with the property that their conditional expected delay is the same, regardless of the time of change. Such rules are called equalizer rules.

Favorable properties of the Shiryaev–Roberts procedure in terms of delay in detection can be shown for the composite alternative hypothesis. However, the road for doing that need not necessarily pass via the sequential theory of optimal stopping. An alternative approach is to produce an information lower bound and deduce the statement.

An information lower bound involves an inequality between the expected delay (or functions thereof), computed under the alternative, and the distribution of false detection that is computed under the null. As one may guess, such inequalities are tightly linked to the concept of a likelihood ratio identity. Indeed, identities like the ones that appear in this chapter can be used to derive such inequalities.

By construction, these inequalities are local in their nature, relating the probability of stopping within a given time interval to the detection delay in the same interval. The criterion of worst case connects together these local statements and produces a global statement that relates the expected time to false alarm with the worst detection delay. Further discussion and the actual development of information lower bounds in the context of sequential change-point detection can be found in [25].

In the context of quickest detection of the emergence of a signal in an image these information lower bounds may be used in order to show that the Shiryaev–Roberts procedure that was analyzed in this chapter is asymptotically optimal if a prior distribution ρ is assumed for the location of the change. This assumption may be relaxed somewhat if the criterion is changed to protection against locations that are hardest to detect in the spirit of the mini-max approach. In such a case a mixture procedure, which is both an optimal procedure with respect to some prior distribution and at the same time an equalizer rule, can be found. Such a procedure is optimal with respect to the mini-max criterion.

The asymptotically optimal procedures are hard to implement. Their construction involves integration of likelihoods, which is a nontrivial task. Instead, one may come up with alternative procedures easier to implement which are almost as efficient. A hint in that direction is given by the approximation that we used for the mixture likelihood on the basis of the Laplace approximation. Proposals in that direction are given in [26]. However, much work is still required in order to turn these vague proposals into practical solutions. We leave the investigation into such important practical issues to the interested reader. Instead, we turn our attention in the next chapter to an unrelated problem that emerges in the context of communication.

9

Buffer overflow

9.1 Introduction

The problem that we analyze in this chapter is the likelihood of a buffer overflow, a classical problem in queuing theory. We will formulate the problem using the language of internet communication. However, the abstract formulation of the problem is relevant to other fields of science such as hydrology, storage and distribution of goods, design of computer architecture, etc.

The problem is simple to state: a communication node is receiving data from a large number of identical sources and transmits the data via an outgoing line of a given bandwidth. The receiving of data from a source may start at a random time and last for a random duration. Occasionally, when a large number of sources send data at the same time, the amount of incoming data may surpass the capacity of the outgoing line. In such a case, the access data is stored in a buffer and awaits its turn to be transmitted. The question to ask is: what should be the size of the buffer?

The answer to the question depends on the characteristics of the system. We will make these characteristics as simple as we can, with just enough difficulty to highlight the issues we want to emphasize.

We assume that the system is examined each quantum unit of time. Without loss of generality we take the quantum unit of time to be equal to 1. Let the bandwidth of the output process be r and let the increment of incoming data between t and $t + 1$ be equal to $\Delta_{t+1} = X_{t+1} - X_t$. Denote by S_t the data in the buffer that awaits transmission at time t. Then the process S_t satisfies the recursion:

$$S_{t+1} = (S_t + \Delta_{t+1} - r)^+ .$$

The storage is empty, making $S_t = 0$, as long as the outgoing line can handle all incoming data. If the demand for communication exceeds the capacity of the

Extremes in Random Fields: A Theory and its Applications, First Edition. Benjamin Yakir.
© 2013 by Higher Education Press. All rights reserved. Published 2013 by John Wiley & Sons, Ltd.

line then S_t becomes positive. In a given instance of time an addition of Δ_{t+1} arrives and a total of r is transmitted out. The difference between the two is the change in the amount of data in the buffer. Once the storage is emptied it stays empty until the beginning of a new accumulation.

Let the size of the buffer be x. A buffer overflow occurs whenever the process S_t exceeds x. Given an interval of time $[0, m]$ we will be interested in the probability of experiencing a buffer overflow:

$$
P\left(\max_{0 \le t \le m} S_t \ge x \right) .
$$

A buffer overflow leads to loss of data and clients' dissatisfaction. Naturally, a system designer is interested in large values of x that make the probability of an overflow small.

The current statement of the event of a buffer overflow is tightly linked to the maximal cusum statistic that was analyzed in Chapter 4. We claim that:

$$
\left\{ \max_{0 \le t \le m} S_t \ge x \right\} = \left\{ \max_{0 \le s \le t \le m} [X_t - X_s - (t-s)r] \ge x \right\} \cup A_0 ,
$$

where A_0 is the event that the process S_t initiated at a positive level and reached the upper threshold x before hitting 0. In order to justify the equality assume that $S_0 = 0$ and consider the integrated process $X_t - tr$. This process evolves in time with a negative drift. Each time a new minimum is obtained all past (negative) accumulation is erased. Between consecutive minima the incremental access from the minimum process is positive and is given by $S_t = (X_t - tr) - \min_{0 \le s \le t}(X_s - sr) = \max_{0 \le s \le t}[X_t - X_s - (t-s)r]$. This incremental access will exceed x if, and only if, there exists an increment larger than x in the process $X_t - rt$. A possibility that is not covered by this description is the case where the process S_t was initiated at a positive level and reached the upper threshold x before hitting 0. This other possibility is covered by the event A_0.

An illustration is given in Figure 9.1. A path of the process $X_t - X_s - (t-s)r$ is plotted and points where positive accumulation initiates are marked. If the process increases after sequential minima then an accumulation is initiated in the buffer. These accumulations, in which the process S_t has a positive value are indicated as solid line intervals. The threshold of overflow is marked for each such interval as a broken line located x units above the level where accumulation initiated. If that level is crossed during an interval of positive accumulation an overflow takes place. An overflow occurred in the last interval.

For an increasing m, the probability of the event A_0 is of smaller order compared with the probability that involves the maximal cusum type of statistic. Henceforth, we will ignore the event A_0 and concentrate only on the probabilistic analysis of the event that is formulated in terms of the maximal cusum statistic.

In many aspects the analysis of the current problem resembles the analysis that was conducted for the cusum change-point detection problem in Chapter 4. However, there are some differences. One difference is that previously the process was

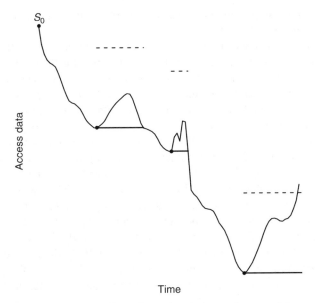

Figure 9.1   The process $X_t - X_s - (t-s)r$. Points where positive accumulation initiates are marked. Solid line intervals indicate regions where the process S_t has a positive value. The threshold of overflow is marked for each such interval as a broken line. An overflow occurs in the last interval.

formulated in terms of log-likelihood ratios which proposed natural candidates for alternative distributions. In the current problem we are given a process and it is for us to associate the process with likelihood ratios. Another important difference will emerge from the construction of the process of incoming data. This construction, which aims at modeling real-life scenarios, will produce a covariance structure different than those encountered before.

In the current application we select the scaling to produce convergence to a Gaussian field. However, due to the specific construction we will get that the resulting limit field resides, in terms of the local behavior of the covariance function, in a domain between the two types of Gaussian field that were considered in the past. These two types emerged, for example, in the context of a Gaussian field of the scanning statistics. One such field was produced when a smooth kernel was used and a different type resulted from the use of an indicator function as the kernel. We specified the difference by a parameter that we denoted by α. That parameter obtained the value 2 for the smooth Gaussian field and the value 1 for the Brownian motion type of field that was produced from a indicator kernel. The parameter that will characterize the limit field in the current case will again be α. But here the parameter will obtain a value strictly between 1 and 2. This type of covariance structure is denoted *long-range dependence*. Of interest will be to compare the results that we will get in the current analysis to results obtained before. Mainly, we would like to identify the role of long-range dependence on the approximation. We will return to this issue in Section 9.4 and in the next chapter.

The stochastic behavior in the current problem originates from the input process X_t. In the initial description we stated that this input is the total accumulation from a large number of identical sources $X_t = \sum_{i=1}^{n} X_{it}$. We add now the simplifying assumption that the summands are stochastically independent. The main complication in the analysis stems from the way in which each such component X_{it} is constructed as an on-off process.

The on-off model describes the process of receiving data from a source as alternating between durations when data are sent – an 'on' period – and durations when the source is idle – an 'off' period. The source alternates back and forth between these two states. The duration in one state is governed by one distribution and the duration in the other state is governed by another distribution. These durations are again assumed to be independent of each other. The process X_{it} corresponds to the total time in the 'on' state between the initial time 0 and the current time t. In particular, if the state was 'on' initially then the total includes the remaining duration of the initiating 'on' period. Likewise, if the process was in the 'on' state at time t then that period is truncated at t when included in the total.

The method by which the model is constructed adds to the complexity of the analysis but, in general, is not sufficient to vary the limit behavior of the local covariance structure. Namely, for typical types of distributions for the duration in states, the distributions that are usually presented in introductory courses in statistics and probability, the produced α will still be equal to 1. The value of the parameter will start to change once the thickness of the tail of the distribution of duration in a state, either 'on', 'off', or both, becomes the dominating factor. In other words, only a heavy tail distribution will lead to a covariance structure characterized by $1 < \alpha < 2$.

A heavy tail distribution, for a positive random variable X, refers to the case where the probability of the tail $P(X > x)$ is asymptotically like $x^{-(1+\beta)}$, for $0 < \beta < 1$. Such a distribution has a finite expectation but an infinite variance, since the integral of x^2 times the density (that behaves in the tail like $x^{-(2+\beta)}$) diverges to infinity. Such a distribution, when used to model the 'on' period, describes demands that may typically be of 'normal' size but occasionally are huge. The load on the system caused by these rare huge demands overwhelms the load produced by standard demands. Historically, it was the empirical observation that the covariance structure in real systems tends to be characterized by a parameter $1 < \alpha < 2$ that motivated the use of heavy tail distribution for modeling and not the other way around.

Interestingly enough, unlike the behavior of a typical 'on' period, the distribution of the initial 'on' period, if the process initiated at the 'on' state, will not even have a finite expectation. The reason is the inspection paradox, a famous phenomenon in renewal theory, which results in a tail probability proportional to $x^{-\beta}$. As it turns out, this super-heavy tail explains the asymptotic derivation and is sufficient to lead to the identification of the parameter of long-range dependence to be equal to $\alpha = 2 - \beta$.

In the next section we give a formal definition of the on-off process and state its properties. The sum of independent on-off processes X_t is rescaled, together with the bandwidth r, in order to produce a process that converges in the limit to a Gaussian process. In Section 9.3 we will carry out the approximation itself. This approximation will involve expressions that depend on α. In Section 9.4 we will discuss the expression that is associated with the constant term that is part of the approximation and present its probabilistic context. In the next chapter the main object is the numerical computation of that constant.

9.2   The statistical model

We concentrate on the process of data received from a given source and drop the subscript i that identifies the source. Later, when we deal with the integrated system, we will reintroduce the subscript. At that point we will give the reader appropriate warning.

9.2.1   The process of demand from a single source

Let $\zeta_1, \zeta_2, \ldots$ be i.i.d. random variables from distribution F_1 and let $\eta_1, \eta_2, \ldots$ be i.i.d. random variables from distribution F_0. An alternating on-off process X_t is produced by taking:

$$
X_t = \begin{cases}
t - \sum_{j=1}^{N_t-1} \eta_j & \text{if } 0 \le t - \sum_{j=1}^{N_t-1} (\zeta_j + \eta_j) < \zeta_{N_t}, \\
\sum_{j=1}^{N_t} \zeta_j & \text{if } \zeta_{N_t} \le t - \sum_{j=1}^{N_t-1} (\zeta_j + \eta_j) < \zeta_{N_t} + \eta_{N_t},
\end{cases}
$$

where N_t is the counting process that is associated with the renewal process S_n formed by the increments $\zeta_j + \eta_j$. The relation between the counting process and the renewal process is given by $\{N_t \le n\} \iff \{S_n > t\}$. A similar definition will initiate the process with an 'off' period η_1 instead of an 'on' period ζ_1.

The alternating on-off process has a stationary distribution that is produced by the examination of $\lim_{s \to \infty}(X_{t+s} - X_s)$ as a function of t. This stationary distribution is the distribution assumed for the process. In particular, the implication of the stationary assumption is that X_t and $(X_{t+s} - X_s)$ have the same distribution, including expectation and variance, for every given t and for all s.

Another way to produce the stationary distribution is to allow the process to initiate either in state 0 with probabilities $E(\eta_j)/[E(\zeta_j) + E(\eta_j)]$ or in state 1 with the complementary probability, and vary the distribution of the initial random variable, η_1 in the first case and ζ_1 in the second.

This alternative description of the stationary distribution results from renewal theory. The probability of being in state 1 at time s is:

$$P\left(s - \sum_{j=1}^{N_s-1} (\zeta_j + \eta_j) < \zeta_{N_s}\right) = \sum_{j=1}^{\infty} \int_0^s P(\zeta_j \geq s - u)P(S_{j-1} = u)du$$

$$= \int_0^s P(\zeta_1 \geq v)h(x - v)dv$$

$$\approx \frac{\int_0^\infty P(\zeta_1 \geq v)dv}{E(\zeta_1 + \eta_1)} = \frac{E(\zeta_1)}{E(\zeta_1) + E(\eta_1)},$$

where h is the density of the renewal measure and the approximation is valid for $s \to \infty$. A similar analysis will give the asymptotic joint distribution of being in state 1 at time s and surviving at this state for at least x units of time:

$$P\left(s + x - \sum_{j=1}^{N_s-1} (\zeta_j + \eta_j) < \zeta_{N_s}\right)$$

$$= \sum_{j=1}^{\infty} \int_0^s P(\zeta_j \geq s + x - u)P(S_{j-1} = u)du$$

$$\approx \frac{\int_x^\infty P(\zeta_1 \geq v)dv}{E(\zeta_1 + \eta_1)} = \frac{\int_x^\infty P(\zeta_1 \geq v)dv}{E(\zeta_1)} \cdot \frac{E(\zeta_1)}{E(\zeta_1) + E(\eta_1)},$$

as required.

The conclusion for an underlying stationary distribution of the process: the conditional distribution of ζ_1, given that the process is at state 1 at the origin, has a density equal to $[1 - F_1(x)]/E(\zeta_1)$, for $0 \leq x < t$, and has a point mass $\int_t^\infty [1 - F_1(u)]du/E(\zeta_1)$ at t. Repeating the same type of analysis with respect to η_1 will give the full description of survival at the initial stage. At the end of the first period the process commences with unaltered distributions at each state. The subsequent state, being the opposite of the initial state, depends on the initial state.

The expectation of the demand that results from the initial state is equal to $\int_0^\infty (u \wedge t)[1 - F_1(u)]du/[E(\zeta_1) + E(\eta_1)]$ and the second moment is $\int_0^\infty (u \wedge t)^2[1 - F_1(u)]du/[E(\zeta_1) + E(\eta_1)]$. If $1 - F_1(t) \sim t^{-(1+\beta)}$, for $0 < \beta < 1$ then the expectation is asymptotic to $t^{1-\beta}$ and the second moment is asymptotic to $t^{2-\beta}$. The variance is also asymptotic to $t^{2-\beta}$ since $2 - \beta > 2(1 - \beta)$ and hence the second moment dominates the square of the first moment. This may give an intuition why one would expect the variance of X_t to behave like $t^{2-\beta} = t^\alpha$.

An actual analysis that proves that this is indeed the asymptotic characteristic of the variance of the source specific demand can be found in [27], together with an asymptotic expression for the variance in a slightly more general case that allows a slowly changing modification of the tail of the distribution. We will not worry about such subtleties and assume henceforth that the variance of the demand is asymptotic to $v^2 t^\alpha$, for some positive constant v and some $1 < \alpha < 2$.

View X_t as the increment from X_0. The assumption that the increments of X_t have a stationary distribution implies that the expectation is a linear function in time. It also implies that the entire covariance structure is determined by the variance of X_t, $t > 0$. Indeed, for any $0 < s < t < \infty$ the variances of X_s, X_t and X_{t-s} are given. The variance of $X_t - X_s$ is equal to the variance of X_{t-s} and is also equal to $\mathrm{Var}(X_t) + \mathrm{Var}(X_s) - 2\mathrm{Cov}(X_t, X_s)$. Therefore, the covariance between X_t and X_s:

$$\mathrm{Cov}(X_t, X_s) = \frac{1}{2}[\mathrm{Var}(X_t) + \mathrm{Var}(X_s) - \mathrm{Var}(X_{t-s})] \qquad (9.1)$$

is computable in terms of the variances of the original process. In particular, for large and far apart t and s the covariance is asymptotic in our case to $0.5v^2[t^\alpha + s^\alpha - (t - s)^\alpha]$.

The location of the origin is arbitrary. Mentally, one can extend the range of definition of the stationary process and include in it also the negative half of the real line. In that case, when both s and t are negative, the covariance between X_s and X_t is a symmetric reflection of the covariance in (9.1) and is computable by substituting s and t by their absolute values. The covariance for $s < 0 < t$ results from the fact that $X_s = -X_{|s|}$. Consequently, the general covariance is obtained by the application of (9.1) to the absolute value of the time parameter and multiplying the outcome by the sign of the product of the time parameters.

9.2.2 The integrated process of demand

It is now time to return to the integrated process of incoming demands for transition of data and reintroduce the index i that specified the source of the demand. Let X_{it} be the on-off process that is associated with source i, $1 \le i \le n$.

The total demand is the sum of the source-specific demands. We would like to examine this process for a large n, having in mind asymptotic approximation that takes n to infinity. This produces a scaling issue. The expected demand grows linearly in n but the variability in the demand growth rate is only proportional to the square root of n. We resolve this issue for the difference between the total demand and the transmission capacity of the outgoing line by scaling differently the random part and the deterministic part of the difference. Specifically, we set $\sqrt{n}\mu = r - \sum_{i=1}^{n} \mathrm{E}(X_{i1})$ and define:

$$Y_t = \frac{1}{\sqrt{n}} \sum_{i=1}^{n} [X_{it} - \mathrm{E}(X_{it})] .$$

Then, after this centering the total demand and rescaling by $\sqrt{n}$, we obtain that the probability that we look to approximate is:

$$P\left(\max_{0 \le t_1 < t_2 \le m} [Y_{t_2} - Y_{t_1} - \mu(t_2 - t_1)] \ge x \right)$$

Notice that we maintained the name x for the threshold although it is now equal to the original threshold, divided by $\sqrt{n}$.

The centered total demand Y_t is a sum of bounded, i.i.d. random variables. It inherits its variance directly from the source specific demand X_{it} since $\mathrm{Var}(Y_t) = \mathrm{Var}(X_{1t})$. It is also stationary. Consequently, the covariance structure of the entire process Y_t is identical to the covariance structure of the process X_t.

The centered demand Y_t, as a sum of i.i.d. random variables, is asymptotically Gaussian. Moreover, the random field $\{Y_{t_1} - Y_{t_2} : 0 \le t_1 < t_2 \le m\}$ converges to a Gaussian random field. By itself, this does not guarantee that the probability we seek to approximate, the probability of exceeding the threshold x, will share the same asymptotic approximation that is derived from the Gaussian limit. The actual approximation depends on the relative rates of divergence to infinity of both x an n. The faster n grows compared with x the closer the limit is to the Gaussian limit and vice versa.

One of our goals in this chapter is to obtain the constant part of the approximation that emerges in the Gaussian setting. These constants are known as Pickands' constants and are the subject of the next chapter. Due to this reason, and due to pure laziness, we will run the asymptotic in a manner that produces the Gaussian limit. Accordingly, whenever we meet a difficult point in the proof we will let n be large enough to wash away that difficulty. At the end, although the large deviation factor will not be replaced by its Gaussian counterpart, the final formula will actually be the formula for the Gaussian fields. This formula has already been obtained in [28] for the case of a Gaussian demand. A more honest analysis will examine more carefully the relation between n and x and will keep track of all important contributions to produce a formula that is applicable in a wider range of scenarios.

9.3   Analysis of statistical properties

The goal in this section is to derive the approximation of the probability of a buffer overflow. This goal will be obtained by the examination of the maximal cusum statistic formed with respect to the process $Y_t - \mu t$. This statistic is produced by the examination of an increment of the process over an interval $(t_1, t_2]$ and the maximization over all such intervals. It is convenient to consider a two-dimensional parameter $\theta = (t_1, t_2)$ and associate the increment of the process Y_t with this parameter: $Y_\theta = Y_{t_2} - Y_{t_1}$. In a similar way, $X_{i\theta} = X_{it_2} - X_{it_1}$, for $i = 1, \ldots, n$, are the increments of demands from the different sources.

9.3.1   The large deviation factor

Let us attempt to identify the large deviation factor and specify the region where the action takes place. Recall the definition of the centered and rescaled total data that are received from the sources during the time interval $\theta = (t_1, t_2)$:

$$Y_\theta = \frac{1}{\sqrt{n}} \sum_{i=1}^{n} \{X_{i\theta} - \mathrm{E}(X_{i\theta})\} ,$$

from which we still need to subtract the access transmission potential of the outgoing line $\mu|\theta|$, for $|\theta| = t_2 - t_1$. Our analysis commences by the consideration of the marginal probabilities $P(Y_\theta - \mu|\theta| \geq x)$ as a function of x, $|\theta|$ and n.

Tilting the distribution of $Y_\theta - E(Y_\theta)$ with the parameter $\xi n^{-\frac{1}{2}}$, while noticing that $\psi'_\theta(\xi) = E_\xi(X_{1\theta}) - E(X_{1\theta})$, gives:

$$P(Y_\theta - \mu|\theta| \geq x) = e^{n[\psi_\theta(\xi n^{-\frac{1}{2}}) - \xi n^{\frac{1}{2}}\psi'_\theta(\xi n^{-\frac{1}{2}})]} E_\xi(e^{-\xi[Y_\theta - E_\xi(Y_\theta)]}; Y_\theta \geq x + \mu|\theta|).$$

The candidate for the large deviation term is the function that multiplies the expectation. For given x and θ, the large deviation term involves the application of the $\xi = \xi_{x,|\theta|,n}$ that solves the equation:

$$x + \mu|\theta| = n^{\frac{1}{2}}\psi'_\theta(\xi n^{-\frac{1}{2}}) \approx \psi''_\theta(0)\xi + \frac{\psi'''_\theta(0)}{2\sqrt{n}}\xi^2 \quad \Rightarrow \quad \xi_{x,\theta,n} \approx \frac{x + \mu|\theta|}{\psi''_\theta(0)}.$$
(9.2)

The approximation results from a first-order Taylor expansion of the function $\psi'_\theta(\xi)$ with respect to ξ. One may obtain a refined approximation that incorporates skewness or validate that the error is small in the first-order approximation by the consideration of a second-order Taylor expansion. The approximation is valid for n increasing fast enough to make the error in the expansion negligible.

Taking a third-order Taylor expansion about 0 of the moment generating function and a second-order expansion of its derivative produces the approximation:

$$n\left[\psi_\theta(\xi n^{-\frac{1}{2}}) - \xi n^{\frac{1}{2}}\psi'_\theta(\xi n^{-\frac{1}{2}})\right] \approx -\frac{1}{2}\psi''_\theta(0)\xi^2 - \frac{1}{3\sqrt{n}}\psi'''_\theta(0)\xi^3.$$
(9.3)

If n is sufficiently large to make the term that involves the third cumulant vanishingly small in comparison with the leading term then we may approximate the exponent of the large deviation factor by plugging in the leading term of (9.3) the approximation from (9.2) of the critical ξ. The resulting exponent is:

$$n\left[\psi_\theta(\xi_{x,\theta,n}n^{-\frac{1}{2}}) - \xi_{x,\theta,n}n^{\frac{1}{2}}\psi'_\theta(\xi_{x,\theta,n}n^{-\frac{1}{2}})\right] \approx -\frac{1}{2}\frac{(x + \mu|\theta|)^2}{\psi''_\theta(0)}.$$

The leading term in the large deviation factor no longer involves n. For a given large value of x, we may consider its behavior as a function of θ in order to identify the exponential decay of the probability to zero. Using the fact that $\psi''_\theta(0) = \text{Var}(X_{i\theta}) \approx v^2|\theta|^\alpha$ we may get that the marginal probability is maximized when the function:

$$x|\theta|^{-\frac{\alpha}{2}} + \mu|\theta|^{1-\frac{\alpha}{2}}$$
(9.4)

is minimized.

Taking a derivative with respect to $|\theta|$ we get that the first derivative is equal to zero when $|\hat{\theta}| = \alpha x/[(2 - \alpha)\mu]$, the derivative is negative to the left of this solution and positive to the right of it. Therefore, the given value of $|\hat{\theta}|$ is indeed the minimum. As a result, the maximum value of the marginal probability is:

$$\log \max_{\theta \in T} P(Y_\theta - \mu|\theta| \geq x) \approx -\frac{x^{2-\alpha}\mu(\alpha)}{(1-\alpha/2)} ,$$

where

$$\mu(\alpha) = \mu^\alpha [2v^2(\alpha/2)^\alpha (1-\alpha/2)^{1-\alpha}] . \tag{9.5}$$

It follows that if the cardinality of T is polynomial in x and if n increases fast enough then:

$$x^{-(2-\alpha)} \log P(\max_{\theta \in T} \{Y_\theta - \mu|\theta|\} \geq x) \to_{x \to \infty} -\mu(\alpha)/(1-\alpha/2) . \tag{9.6}$$

It is interesting to compare the given rate to the rate obtained in the case $\alpha = 1$. This case is associated with short-range dependence. In such a case we get that the exponent of x is equal to $2 - \alpha = 1$ and the constant that multiplies x is equal to $-\mu(\alpha)/(1-\alpha/2) = -2\mu/v^2$.

9.3.2 Preliminary localization

In order to identify intervals θ that make a non-negligible contribution we may take a closer look at the marginal probabilities. The intervals of length $x\alpha/[(2-\alpha)\mu]$ form a ridge in the surface of marginal survival functions. A first-order upper bound on the probability of a buffer overflow, which misses only in the polynomial and constant terms, is the sum of the marginal probabilities. This integral is dominated by the probabilities in the vicinity of the ridge. We want to identify the relevant neighborhood about the ridge, with the intention of eliminating intervals that make a negligible contribution to the probability of an overflow.

Specifically, we examine the rate of decrease of the marginal probabilities as a function of the deviation of the length of the interval from the maximizing length. Indeed, the ratio between a marginal probability and the maximal marginal probability is dominated by the difference between the terms that go into the exponent. The first derivative of the function (9.4) was used in order to find the maximizing interval length. The second derivative, evaluated at the maximizer, is:

$$\frac{\alpha(\alpha+2)x}{4|\hat{\theta}|^{\frac{\alpha}{2}+2}} - \frac{(2-\alpha)\alpha\mu}{4|\hat{\theta}|^{\frac{\alpha}{2}+1}} = \frac{[(1-\alpha/2)\mu]^{\frac{\alpha}{2}+2}}{[x\alpha/2]^{\frac{\alpha}{2}+1}} .$$

Consequently, a Taylor expansion of the function in (9.4) produces:

$$x|\theta|^{-\frac{\alpha}{2}} + \mu|\theta|^{1-\frac{\alpha}{2}} \approx \frac{x^{1-\frac{\alpha}{2}}\mu^{\frac{\alpha}{2}}}{(\alpha/2)^{\frac{\alpha}{2}}(1-\alpha/2)^{1-\frac{\alpha}{2}}} + \frac{(|\theta|-|\hat{\theta}|)^2}{2} \frac{[(1-\alpha/2)\mu]^{\frac{\alpha}{2}+2}}{[x\alpha/2]^{\frac{\alpha}{2}+1}} .$$

Taking the square of the approximation, divided by $2v^2$, and ignoring smaller order terms we get:

$$\frac{(x|\theta|^{-\frac{\alpha}{2}} + \mu|\theta|^{1-\frac{\alpha}{2}})^2}{2v^2} \approx \frac{x^{2-\alpha}\mu(\alpha)}{(1-\alpha/2)} + \left(\frac{|\theta| - |\hat{\theta}|}{x^{\frac{\alpha}{2}}}\right)^2 \frac{\mu^{\alpha+2}(1-\alpha/2)^{\alpha+1}}{2v^2(\alpha/2)^{\alpha+1}} .$$

We may conclude that the only intervals that contribute non-negligibly to the approximation are those with lengths in the range $|\hat{\theta}| \pm O(x^{\frac{\alpha}{2}})$. Henceforth, we restrict T to such intervals.

Compare this with the situation where short-range dependence was involved. The width of the region in that case was proportional to $x^{\frac{1}{2}}$.

9.3.3   Approximation by a cruder grid

The next step is the approximation of the collection T by a sub-collection $\hat{T} \subset T$, which is on the one hand sparse so the local field converges to a process with a discrete time domain but is also dense enough to assure that the approximation produced by restricting maximization to the sub-collection is accurate.

We mimic the argument that was used in the example of a simple Gaussian scanning statistic in Chapter 3. The approximation of the target probability by the probability that uses maximization only over the subset $\hat{T}$ involves an obvious lower bound and an upper bound. The upper bound is produced by the consideration of local regions, denoted T_θ, that are associated with elements of $\hat{T}$.

The probabilities that should be analyzed in the construction of the upper bound are of the type:

$$P(Y_\theta - \mu|\theta| \le x - \epsilon x^{\alpha-1}, \max_{\vartheta \in T_\theta}(Y_\vartheta - Y_\theta) \ge (x - Y_\theta + \mu|\theta|) \vee (\epsilon x^{1-\alpha})) ,$$

(9.7)

where we arrange the subsets T_θ so that $s_1 \le t_1 \le t_2 \le s_2$, for all $\vartheta = (s_1, s_2) \in T_\theta$. The largest standard deviation of an increment $Y_\vartheta - Y_\theta$ should not exceed $x^{1-\alpha}$ (times a small quantity), which suggests using $\delta x^{2(1-\frac{1}{\alpha})}$ as the the maximal value of $|s_1 - t_1| \vee |s_2 - t_2|$.

The key element in obtaining a probabilistic bound on the probability in (9.7) is the identification of a bound on the tail of increments of the local field. In the Gaussian case the tail of the standardized increments was the Gaussian tail, which converges to zero faster than exponential. This fact, together with a chaining argument, produces the proof of Fernique's inequality that we used to produce a discrete approximation of the parameter space in the case of a Gaussian scanning statistic. We follow similar steps in the current argument. Consequently, we may want also to investigate the tail behavior of increments such as $Y_{t_2+s} - Y_{t_2+r}$, for $0 < s < r \le \delta x^{2(1-\frac{1}{\alpha})}$ on one side of θ and similar increments on the other side.

The original Gaussian argument included conditioning on the value of the random variable that corresponds to θ. The density of that random variable produced the large deviation factor. Let us do something similar here. Denote

$A_j = \{-j - 1 \le Y_\theta - \mu|\theta| - x \le -j\}$. An upper bound on the joint distribution of the random variable associated with θ and the maximum of the local field may be obtained via:

$$2 \sum_{j=\epsilon x^{\alpha-1}}^{Cx^{\frac{\alpha}{2}}} P\left(A_j, \max_{0 \le s \le \delta x^{2(1-1/\alpha)}} |Y_{t_2+s} - Y_{t_2}| \ge j\right), \qquad (9.8)$$

for some large C. The truncation of the upper limit in the summation is justified by the same type of considerations that lead to restriction of the parameter set to the range where action occurs.

Denote by $\mathcal{F}_n = \sigma\{X_{i\theta} : 1 \le i \le n\}$ the σ-algebra generated by demands from the different sources during the time interval θ. We analyze the probabilities in the sum by conditioning on the given σ-algebra. The events A_j are measurable with respect to the σ-algebra. The event that involves maximization is analyzed in the conditional distribution and it is bounded by two sums. One is a sum of indicators of events that are measurable with respect to the σ-field. The other is a sum of bounds on the conditional probability of the event of maximization. These probabilistic bounds hold on the complementary of the events that formed the sum of indicators. The resulting bound is our substitute for Fernique's inequality.

Let $\Lambda_n = \Lambda_n(r, s) = E(Y_{t_2+s} - Y_{t_2+r}|\mathcal{F}_n)$ be the conditional expectation of a (scaled) demand in an increment of time and let $\bar{\sigma}_n^2 = \bar{\sigma}_n^2(r, s) = \mathrm{Var}(Y_{t_2+s} - Y_{t_2+r}|\mathcal{F}_n)$ be the conditional variance during that period. Define:

$$M_n = M_n(r, s) = \max_{1 \le i \le n} \{(r - s) - E(X_{i(t_2+s)} - X_{i(t_2+r)}|X_{i\theta})\} n^{-\frac{1}{2}} \le (r-s)n^{-\frac{1}{2}}.$$

As a conclusion of Theorem A.4 we get that:

$$P(Y_{t_2+s} - Y_{t_2+r} \ge y|\mathcal{F}_n) \le \exp\left\{-\frac{1}{2\bar{\sigma}_n^2}(y - \Lambda_n)^2 \frac{1}{1 + (y - \Lambda_n)M_n/(3\bar{\sigma}_n^2)}\right\},$$

for $y \ge \Lambda_n$. For a small M_n, the right-hand side has a normal-type behavior and a function of $z = (y - \Lambda_n)/\sigma_n$, when z is not too large and a Poisson-type behavior when z is very large. We use this upper bound to produce a chaining argument for bounding the maxima of the local field.

A chaining argument, like the one considered in Chapter 6, involves the formation of a pedigree of partitions of the parameter space. Each offspring partition is a refinement of the predecessor partition and may be produced, for example, by partitioning each subset into two new subsets. For a given partition, the central element in each subset is marked. The distance between the central element in a partition and a central element in the refined partition is half the radius of the set. One may construct increments. For each parental central element increments are constructed, one for each of its two offspring in the refined partition. The number of such increments in a generation is the number of subsets in the refined partition.

This construction can be described by a binary tree. The leaves of the tree are the original collection of elements of the local field. The root of the tree is the element of the field with central index. Each leaf is connected to the root by a branch of length equal to the depth of the tree. One may associate a branch with a series of increments between a parental central element and one of its central offspring. Thereby, each element in the local field is connected to the root by such a sequence of increments.

We want to bound the probability that there exists a leaf that is larger in value than some quantity. If we allocate to each generation a part of the quantity, then in order to exceed the quantity at least one generation should have a central member that exceeded its parent by more than the allocated amount. Consequently, we may bound the probability of the event in question, an event that is produced by a union of events that involves an offspring that surpasses its parent, by the sum of probabilities of the events in the union.

Formally, let the number of elements on the right-hand side of the local field be equal to $u = \delta x^{2(1-\frac{1}{\alpha})}$ and let k, $0 \leq k \leq \log_2 u$ be the index of the generation. The number of central increments in the kth generation is 2^k and the distance between the central elements and their parent is $u2^{-k}$. Let $j \geq \sum_{k=0}^{\log_2 u} j p_k$, where $p_k = 1/[(k+1)(k+2)]$ are positive numbers with a sum bounded by 1. Then

$$P\left(\max_{1 \leq s \leq m} Y_{t_2+s} \geq j | \mathcal{F}_n\right) \leq \sum_{k=0}^{\log_2 u} \sum_{i=1}^{2^k} P(|Y_{c_{i,k}} - Y_{p(c_{i,k})}| \geq j p_k | \mathcal{F}_n),$$

where $p(c_{i,k})$ is the central parent of the central element $c_{i,k}$.

Let $\Lambda_{i,k} = \Lambda_n(c_{i,k}, p(c_{i,k}))$, $\overline{\sigma}_{i,k}^2 = \overline{\sigma}_n^2(c_{i,k}, p(c_{i,k}))$ and define $\sigma_k^2 = \text{Var}(X_{m2^{-k}}) \approx v^2 \delta^\alpha x^{2(\alpha-1)} 2^{-\alpha k}$. Define:

$$E_{j,i,k} = \{|\Lambda_{i,k}| \geq 0.25 j p_k\} \cup \{\overline{\sigma}_{i,k}^2 > 1.5 \sigma_k^2\},$$

and let n be large enough to assure that the normal and not the Poisson part of the upper bound rules. Then:

$$P\left(\max_{1 \leq s \leq m} Y_{t_2+s} \geq j | \mathcal{F}_n\right) \leq \sum_{k=0}^{\log_2 u} \exp\left\{(k+1)\log 2 - \frac{1}{24}\frac{j^2 p_k^2}{\sigma_k^2}\right\} + \sum_{k=0}^{\log_2 u} \sum_{i=1}^{2^k} 1_{E_{j,i,k}}.$$

This bound, composed of a sum that emerged from the application of the concentration inequality and a sum of events measurable with respect to $\mathcal{F}_n$, will be used to evaluate (9.8).

By tilting the distribution of Y_θ and applying a bound on the density of the random variable under the alternative we can get, for a large n, a bound on the probability of the event A_j. This bound, together with the sum that resulted from the concentration inequality gives:

$$
\sum_{j=\epsilon x^{\alpha-1}}^{Cx^{\frac{\alpha}{2}}} \sum_{k=0}^{\log_2 u} \frac{c}{\sqrt{2\pi \, \psi_\theta''(0)}} \exp\left\{ -\frac{1}{2} \frac{(x+\mu|\theta|-j-1)^2}{\psi_\theta''(0)} + (k+1)\log 2 - \frac{1}{24} \frac{j^2 p_k^2}{\sigma_k^2} \right\}.
$$

Pulling $(2\pi \, \psi_\theta''(0))^{-\frac{1}{2}} \exp\{-\frac{1}{2}(x+\mu|\theta|)^2/\psi_\theta''(0)\}$ in front of the summation will leave a sum that is bounded by:

$$
\sum_{j=\epsilon x^{\alpha-1}}^{Cx^{\frac{\alpha}{2}}} \sum_{k=0}^{\log_2 u} \exp\left\{ z(j-1) + (k+1)\log 2 - \frac{1}{24} \frac{2^{\alpha k}(p_{k,j})^2}{v^2 \delta^\alpha} \left(\frac{j}{x^{\alpha-1}} \right)^2 \right\},
$$

for some bounded number z. This double-sum can be approximated by a sum with respect to k of integrals with respect to j. Changing the variable to $y = jx^{1-\alpha}$ and carrying out the integration with respect to y will produce a sum in k of rapidly converging Gaussian survival functions. Each of these functions will have the term $\epsilon/\delta^{\frac{\alpha}{2}}$ in the exponent.

For the final play in this part of the argument we may note that the change in variable produces the factor $x^{\alpha-1}$. This factor is then absorbed by the change of variable when an integration with respect to $\theta \in \hat{T}$ takes place. The term that results from the final integration with respect to θ is of the order of magnitude of the probability that is obtained by restricting the maximization to $\hat{T}$, modified by terms that depend on δ. Just as in the case of the scanning statistic, the modification includes a term that grows linearly with the decrease of δ and a term that decreases exponentially fast. The exponential term wins.

In order to complete the justification of the use of the proposed grid we still need to bound the sum of indicators that was left out. This corresponds to the analysis of the sum of probabilities:

$$
\sum_{j=\epsilon x^{\alpha-1}}^{Cx^{\frac{\alpha}{2}}} \sum_{k=0}^{\log_2 u} \sum_{i=1}^{2^k} P(A_j \cap E_{j,i,k}).
$$

Each of the probabilities in the sum can be bounded by:

$$
P(A_j \cap E_{j,i,k}) \le P(A_j \cap \{|\Lambda_{i,k}| \ge 0.25 j p_k\}) + P(\overline{\sigma}_{i,k}^2 > 1.5\sigma_k^2).
$$

The second probability involves an average of independent and bounded random variables with expectation strictly less than σ_k^2. This probability is exponentially small as a function of n and is negligible if n is diverging fast enough to infinity.

For the bounding of the first probability we may use Theorem 5.3, at least for values of j and k that are relatively small. The local limit is applied, after tilting, to the random variable Y_θ that defines the event A_j. The random variable $\Lambda_{i,k}$

has an asymptotic zero mean and it converges to a normal random variable. The variance is smaller than σ_k^2. As we will prove later on, the two random variables are asymptotically uncorrelated. Again, by assuring that n is large enough we may use the Gaussian limit and produce a bound that involves the density of Y_θ times a rapidly decreasing function of j and k. The argument that was used before in the context of a concentration inequality can be repeated to produce an overall sum that converges to zero with the decrease of δ.

This completes the proof of the validity of the approximation of T by $\hat{T}$. Henceforth we will consider the sparse set to be the set of parameters and rename it, accordingly, T.

9.3.4 Measure transformation

The preparations took time and effort but now we are ready for the application of the likelihood ratio identity. Natural candidates for log-likelihoods are obtained by tilting: $\ell_\theta = \xi Y_\theta - n\psi_\theta(\xi n^{-\frac{1}{2}})$. But which value of the tilting parameter ξ should we use? The answer to this question is not unique and any reasonable selection may show advantage in some stages of the proof but be less favorable in other stages. Here we propose to use for all $\theta \in T$ the value of ξ that is associated with the maximizing parameter $\hat{\theta}$. This gives:

$$\xi = \xi_{x,\theta,n} \approx \frac{x + \mu|\hat{\theta}|}{\psi_\theta''(0)} \approx \frac{x^{1-\alpha}\mu^\alpha}{v^2(\alpha/2)^\alpha(1-\alpha/2)^{1-\alpha}} = 2\mu(\alpha)x^{1-\alpha} .$$

In the following we use this value of ξ. This value depends on x, but is independent of θ.

We seek to approximate the probability of an event A that can be rewritten in terms of the given log-likelihood ratios:

$$A = \left\{ \max_{\theta \in T}[\ell_\theta + n\psi_\theta(\xi n^{-\frac{1}{2}}) - \xi\mu|\theta|] \geq \xi x \right\} = \left\{ \max_{\theta \in T}[\ell_\theta + k(|\theta|)] \geq \xi x \right\} ,$$

where

$$k(|\theta|) = n\psi_\theta(\xi n^{-\frac{1}{2}}) - \xi\mu|\theta| .$$

For the likelihood ratio identity we use $\sum_{\theta \in T} \exp\{\ell_\theta + k(|\theta|)\}$. The resulting identity produces the representation:

$$\begin{aligned}
P(A) &= \sum_{\theta \in T} e^{k(|\theta|)} E_\theta \left[\frac{1}{\sum_{\vartheta \in T} e^{\ell_\vartheta + k(|\vartheta|)}}; A \right] \\
&= \sum_{\theta \in T} e^{k(|\theta|)} E_\theta \left[\frac{e^{-(\ell_\theta + k(|\theta|))}}{\sum_{\vartheta \in T} e^{\ell_\vartheta - \ell_\theta + k(|\vartheta|) - k(|\theta|)}}; A \right] \\
&= \sum_{\theta \in T} e^{k(|\theta|) - \xi x} E_\theta \left[\frac{M_\theta}{S_\theta} e^{-(\tilde{\ell}_\theta + m_\theta)}; \tilde{\ell}_\theta + m_t \geq 0 \right] , \quad (9.9)
\end{aligned}$$

for

$$S_\theta = \sum_{\vartheta \in T} e^{\ell_\vartheta - \ell_\theta + k(|\vartheta|) - k(|\theta|)} \ , \qquad M_\theta = \max_{\vartheta \in T} e^{\ell_\vartheta - \ell_\theta + k(|\vartheta|) - k(|\theta|)} \ ,$$

$m_\theta = \log M_\theta$, and $\tilde{\ell}_\theta = \ell_\theta + k(|\theta|) - \xi x$.

9.3.5   The localization theorem

Unlike previous applications the representation is a weighted sum instead of a regular sum. This should not discourage us from proceeding as usual with the application of the localization theorem. This application involves the global term $\tilde{\ell}_\theta$ and the local random field $\{w_\theta(\vartheta) : \vartheta \in T\}$, where $w_\theta(\vartheta) = \ell_\vartheta - \ell_\theta + k(|\vartheta|) - k(|\theta|)$.

Our next goal is to identify the limit distribution of the global term and of the elements in the local field in the vicinity of θ, first as a limit in n and then as a limit in x.

From the definition of the terms involved we get that:

$$w_\theta(\vartheta) = \xi(Y_\vartheta - Y_\theta) - (|\vartheta| - |\theta|)\xi\mu \ .$$

The limit distribution of the elements of the local field is Gaussian and will be determined by the asymptotic expectation of each element, the asymptotic variance, and the covariance between elements, all computed under the tilted P_θ-distribution. It is useful to note that the demands from the different sources $X_{i\vartheta} - X_{i\theta}$, $1 \le i \le n$, are independent not only under the null distribution but also under the tilted alternative.

The P_θ-expectation of a specific demand $X_{i\vartheta} - X_{i\theta}$ is equal to:

$$E((X_{i\vartheta} - X_{i\theta})e^{\xi X_{i\theta} n^{-\frac{1}{2}} - \psi_\theta(\xi n^{-\frac{1}{2}})})$$

and for large n this is asymptotic to:

$$E(X_{i\vartheta} - X_{i\theta}) + \xi n^{-\frac{1}{2}}\mathrm{Cov}(X_{i\vartheta} - X_{i\theta}, X_{i\theta}) \ .$$

By stationarity,

$$\mathrm{Cov}(X_{i\vartheta} - X_{i\theta}, X_{i\theta}) = \frac{1}{2}\{\mathrm{Var}(X_{i\vartheta}) - \mathrm{Var}(X_{i\theta}) - \mathrm{Var}(X_{i\vartheta} - X_{i\theta})\} \ .$$

Consequently,

$$E_\theta[w_\theta(\vartheta)] = -\frac{\xi^2}{2}\mathrm{Var}(X_{i\vartheta} - X_{i\theta}) + \frac{\xi^2}{2}[\mathrm{Var}(X_{i\vartheta}) - \mathrm{Var}(X_{i\theta})] - (|\vartheta| - |\theta|)\xi\mu \ .$$

Moreover, up to a $o(x^{\alpha-1})$ term,

$$\frac{\xi}{2}[\text{Var}(X_{i\vartheta}) - \text{Var}(X_{i\theta})]$$

$$= \frac{v^2|\theta|^{\alpha-1}\xi}{2}|\theta|[(|\vartheta|/|\theta|)^{\alpha} - 1]$$

$$= \left(1 - \frac{|\theta| - |\hat{\theta}|}{|\hat{\theta}|}\right)^{\alpha-1} \times \frac{\mu|\theta|}{\alpha}\left[\left(1 + \frac{|\vartheta| - |\theta|}{|\theta|}\right)^{\alpha} - 1\right]$$

$$= \left(1 - \frac{|\theta| - |\hat{\theta}|}{|\hat{\theta}|}\right)^{\alpha-1} \times \left(1 + \frac{|\vartheta^*| - |\theta|}{|\theta|}\right)^{\alpha-1} \times (|\vartheta| - |\theta|)\mu$$

$$= (|\vartheta| - |\theta|)\mu + O\left(||\theta| - |\vartheta||\, x^{-(1-\frac{\alpha}{2})}\right).$$

The last equation follows from the fact that length $|\vartheta^*|$ belongs to the interval of lengths that are between $|\vartheta|$ and $|\theta|$ and all the lengths are $|\hat{\theta}| \pm O(x^{\frac{\alpha}{2}})$.

One may conclude that:

$$E_{\theta}[w_{\theta}(\vartheta)] = -\frac{\xi^2}{2}\text{Var}(X_{1\vartheta} - X_{1\theta}) + O(||\theta| - |\vartheta||x^{-\frac{\alpha}{2}}) + o(1). \quad (9.10)$$

Comparing the first term in the expectation to the $O(\cdot)$ term one may see that the latter term, in the worst case scenario, is of the order of the square root of the former term.

The variance under the alternative of an increment of the local field converges to the null variance:

$$\text{Var}_{\theta}[w_{\theta}(\vartheta) - w_{\theta}(\eta)] \to_{n \to \infty} \xi^2 \text{Var}(X_{1\vartheta} - X_{1\eta}). \quad (9.11)$$

To conclude, the covariance structure of the limit Gaussian field is produced by stationary increments. Moreover, for values in the vicinity of θ, the variance of an element is asymptotic to twice the negative expectation of the element.

The global term corresponds to the random variable $\tilde{\ell}_{\theta} = \ell_{\theta} + k(|\theta|) - \xi x$. The limit P_{θ}-distribution of this term, as a sum of i.i.d. random variables, is normal. The asymptotic expectation is equal to $E_{\theta}(\tilde{\ell}_{\theta}) \approx \xi^2 \text{Var}(X_{1\theta}) - \xi\mu|\theta| - \xi x$. This expectation is vanishingly small for all θ on the ridge $|\theta| = |\hat{\theta}|$ and is of order $O(x^{1-\frac{\alpha}{2}})$ for other interval lengths in T. The limit variance is $\text{Var}_{\theta}(\tilde{\ell}_{\theta}) \approx \xi^2 \text{Var}(X_{1\theta})$. Hence, for large values of x, the variance is of magnitude:

$$\xi^2 \text{Var}(X_{1\theta}) \approx x^{2-\alpha}v(|\theta|/x)^{\alpha}[2\mu(\alpha)]^2 \approx \frac{x^{2-\alpha}\mu^{\alpha}}{v^2(\alpha/2)^{\alpha}(1 - \alpha/2)^{2-\alpha}} = \sigma^2 x^{2-\alpha}.$$

$$(9.12)$$

We examined the limit as $n \to \infty$ for the local field and did not pay close attention to the role played by the threshold. This will be fixed with the application of the localization theorem. Set $\theta = (j_1 \delta x^{2(1-\frac{1}{\alpha})}, j_2 \delta x^{2(1-\frac{1}{\alpha})}) \in T$. For the running parameter in the asymptotic derivation one may use $\kappa = x^{2-\alpha}$. The local σ-algebra $\hat{\mathcal{F}}_\kappa$ is generated by a finite collection of increments $w_\theta(\vartheta)$. The increments belong to the sub-collection:

$$T_\theta = \{\vartheta = (\vartheta_1, \vartheta_2) : \vartheta_1 = i_1 \delta x^{2(1-\frac{1}{\alpha})}, \ \vartheta_2 = i_2 \delta x^{2(1-\frac{1}{\alpha})}, \ |i_1 - j_1| \vee |i_2 - j_2| \leq \tau\},$$

for some large enough but fixed τ.

The variance of $w_\theta(\vartheta)$ is asymptotically proportional to the variance of $X_{1\vartheta} - X_{1\theta}$. The latter variance corresponds to the variance of the demand in the increment $[i_1, j_1) \cdot \delta x^{2(1-\frac{1}{\alpha})}$, the variance in the increment $[i_2, j_2) \cdot \delta x^{2(1-\frac{1}{\alpha})}$, and the covariance between the demands in these intervals. The distance between the two sub-intervals is of order $|\hat{\theta}|$, which is proportional to x. With the increase in κ the variances converge to finite positive values and the covariance vanishes:

$$\lim_{x \to \infty} \text{Var}_\theta[w_\theta(\vartheta)] = \sigma_\delta^2\{|i_1 - j_1|^\alpha + |i_2 - j_2|^\alpha\},$$

for $\sigma_\delta^2 = 4v^2[\mu(\alpha)]^2 \delta^\alpha$. A similar argument applied to increments $w_\theta(\vartheta) - w_\theta(\eta)$, together with the assumed stationarity, will produces for the covariance:

$$\lim_{x \to \infty} \text{Cov}_\theta[w_\theta(\eta), w_\theta(\vartheta)] = \sigma_\delta^2\{R(h_1 - j_1, i_1 - j_1) + R(h_2 - j_2, i_2 - j_2)\},$$

where

$$R(h, i) = \frac{1}{2}\{|h|^\alpha + |i|^\alpha - |h - i|^\alpha\},$$

multiplied by the sign of $(h \cdot i)$. It follows that the limiting covariance structure of the elements of the field $\hat{\mathcal{F}}_\kappa$ corresponds to a sum of two independent and stationary Gaussian processes, each with a correlation structure characterized by the function R.

For the limit of the expectations consult (9.10). The first term in the limit with respect to n converges to negative one half of the variance. The second, the big $O(\cdot)$ term, is bounded by a constant times $x^{2-2/\alpha-\alpha/2}$. This term converges to zero since $2 - 2/\alpha - \alpha/2 < 0$, for $1 < \alpha < 2$. The third term in the approximation of the expectation converges to zero. It follows that the limit expectation in both independent Gaussian processes, one for each end of the interval θ, is equal to $-0.5\,\sigma_\delta^2|i - j|^\alpha$, where $i - j$ stands for $i_1 - j_1$ or $i_2 - j_2$, depending on which end one considers.

The term $E[\hat{\mathcal{M}}/\hat{\mathcal{S}}]$ is the expectation of the ratio between the maximum and the sum of the exponentiated limit field, restricted to the subset T_θ. This limit field can be represented as a sum of two i.i.d. processes, one indexed by i_1 and the other by i_2. The result is that the ratio is equal to the product of two ratios, one for each process. A corollary of independence between the processes is that the expectation of the product of ratios is equal to the product of the expectations, namely the square of a single expectation.

Abusing the notation in Theorem 5.2, let us denote by $E[\mathcal{M}/\mathcal{S}]$ the expectation of a ratio applied to an infinite double-ended one-dimensional process:

$$E[\mathcal{M}/\mathcal{S}] = E\left[\frac{\max_i e^{\sigma_\delta W_i - 0.5\sigma_\delta^2 |i|^\alpha}}{\sum_i e^{\sigma_\delta W_i - 0.5\sigma_\delta^2 |i|^\alpha}}\right], \qquad (9.13)$$

for $\sigma_\delta^2 = 4v^2[\mu(\alpha)]^2\delta^\alpha$ and for $\{W_i : -\infty < i < \infty\}$ a stationary Gaussian process with a covariance structure characterized by $\mathrm{Var}(W_i) = |i|^\alpha$. We get that $E[\hat{\mathcal{M}}/\hat{\mathcal{S}}] \to \{E[\mathcal{M}/\mathcal{S}]\}^2$, as $\tau \to \infty$. This identifies for us the constant that should be used in the conclusion of Theorem 5.2. Similarly, the component that is associated in the theorem with the global term should appear in the conclusion as the evaluation in 0 of the limit normal density.

9.3.6 Integration

We return to the evaluation, following the application of the localization theorem, of the probability of an overflow. The final step in the approximation of the probability results from replacing the expectations in the representation (9.9) by their approximations that are produced by Theorem 5.2. Subsequently, the sum is approximated by an integral.

First, let us identify the exponent that is produced from the combination of the weights and the normal density that is associated with the global term. The exponent of the weights is of the form $k(|t|) - \theta x$. The localization theorem contributes to the exponent the negative square of the expectation of the global term $\tilde{\ell}$, divided by twice its variance. Consequently, the combined exponent becomes:

$$k(|\theta|) - \xi x - \frac{[E_\theta(\tilde{\ell})]^2}{2\mathrm{Var}_\theta(\tilde{\ell})} \approx k(|\theta|) - \xi x - \frac{(\xi^2\mathrm{Var}(X_{1\theta})/2 + k(|\theta|) - \xi x)^2}{2\xi^2\mathrm{Var}(X_{1\theta})}$$

$$\approx -\frac{(\xi^2\mathrm{Var}(X_{1\theta}))/2 - k(|\theta|) + \xi x)^2}{2\xi^2\mathrm{Var}(X_{1\theta})}$$

$$\approx -\frac{(\mu|\theta| + x)^2}{2\mathrm{Var}(X_{1\theta})} \approx -\frac{(\mu|\theta|^{1-\frac{\alpha}{2}} + x|\theta|^{-\frac{\alpha}{2}})^2}{2v^2},$$

with the initial approximations resulting form an increase in n, gradually replaced by approximations that result from an increase in x.

The previous analysis that led to restricting the parameter set T to a region about the ridge of marginal probabilities is relevant in the current assessment of the integration step. Specifically, when we combine the conclusion of the localization theorem and the current combination of the weight and the density function of the global term, and include it all in representation (9.9), we get:

$$P\left(\max_{\theta \in T}(Y_\theta - \mu|\theta|) \geq x\right)$$

$$= \sum_{\theta \in T} e^{k(|\theta|)-\xi x} E_\theta \left[\frac{M_\theta}{S_\theta} e^{-(\tilde{\ell}_\theta + m_\theta)}; \tilde{\ell}_\theta + m_\theta \geq 0\right]$$

$$\approx e^{-\frac{x^{2-\alpha}\mu(\alpha)}{(1-\alpha/2)}} [E(\mathcal{M}/\mathcal{S})]^2 \frac{1}{\sqrt{2\pi} x^{1-\frac{\alpha}{2}}\sigma} \sum_{\theta \in T} e^{-(|\theta|-|\hat{\theta}|)^2 x^{-\alpha} \frac{\mu^{\alpha+2}(1-\alpha/2)^{\alpha+1}}{2v^2(\alpha/2)^{\alpha+1}}}$$

$$\approx e^{-\frac{x^{2-\alpha}\mu(\alpha)}{(1-\alpha/2)}} [E(\mathcal{M}/\mathcal{S})]^2 \frac{\varsigma}{x^{1-\frac{\alpha}{2}}\sigma} \sum_{i=1}^{m/[\delta x^{2(1-\frac{1}{\alpha})}]} \sum_{|j-\hat{j}|\leq C/\varsigma} \frac{\exp\left\{-\frac{(j-\hat{j})^2}{2\varsigma^2}\right\}}{\sqrt{2\pi}\varsigma}$$

$$\approx m e^{-\frac{x^{2-\alpha}\mu(\alpha)}{(1-\alpha/2)}} x^{\frac{4}{\alpha}+\alpha-5} \delta^{-2} [E(\mathcal{M}/\mathcal{S})]^2 \frac{v^2(\alpha/2)^{\alpha+\frac{1}{2}}}{\mu^{\alpha+1}(1-\alpha/2)^{\alpha-\frac{1}{2}}} .$$

The last expression is the approximation for the probability of a buffer overflow, not in the original formulation that dealt with quantum units of time but, instead, when the examination of the system is in time units of size $\delta x^{2(1-1/\alpha)}$. We used in the derivation the auxiliary notation:

$$\varsigma^{-1} = v^{-1}\delta x^{2-\frac{2}{\alpha}-\frac{\alpha}{2}} \mu^{\frac{\alpha+2}{2}} (2/\alpha - 1)^{\frac{\alpha+1}{2}}$$

and σ^2 that is defined in (9.12).

The expression is obtained after letting τ from the localization theorem go to infinity and allowing the restriction on the range about the ridge of the values that belong to T to be lifted. In the next section we will allow δ to go to zero and obtain an approximation for the probability of an overflow in the original problem.

Note that for the case $\alpha = 1$ there is no polynomial term in x and the last constant is equal to $v^2/[2\mu^2]$.

9.3.7 Checking the conditions of the localization theorem

Before we rush to publish the approximation we may want to check that the conditions of Theorem 5.2 and Theorem 5.3 are satisfied. There are five such conditions for the former and a requirement of asymptotic zero correlation for the latter.

Consider Theorem 5.2. Condition I* trivially holds and Condition III* had, in practice, already been discussed.

9.3.8 Checking Condition IV*

Condition IV* is basically a corollary of Theorem 5.3, but in order to use that theorem we still need to establish that the global term and the local field are asymptotically uncorrelated. To prove this claim we take n and x to infinity simultaneously. The covariance between a local term $w_\theta(\vartheta)$ and the global term $\tilde{\ell}_\theta$ is proportional to the null covariance between $X_{1\vartheta} - X_{1\theta}$ and $X_{1\theta}$. That covariance was analyzed before when we considered the expectation of $w_\theta(\vartheta)$ under the alternative. The leading term in the covariance stems from the difference $\|\vartheta| - |\theta\|$. This difference is bounded, over the local region T_θ, by a constant times $x^{3-\alpha-\frac{2}{\alpha}}$. This quantity should still be divided by the standard deviation of the global term, which is of order of magnitude $x^{1-\frac{\alpha}{2}}$, and the standard deviation of the local element, which is of the order of a constant. The resulting correlation is of the order $x^{2-\frac{\alpha}{2}-\frac{2}{\alpha}}$ and converges to zero. This is the case since the exponent is negative.

9.3.9   Checking Condition V*

For Condition V* we may note that $M_\kappa \geq \hat{M}_\kappa \geq 1$ so only the tail of M_κ needs to be controlled. A bound on this term may be obtained by an application of the concentration inequality given in Theorem A.4. We may use the expression for the expectation of an increment of the local field that is given in (9.10), the expression for the variance that is given in (9.11, with $\eta = \theta$), and the fact that the increment is a sum of independent and bounded random variables. The expectation is dominated by the first ingredient and hence negative. When the Hamming distance between ϑ and θ is more than $c \log x$ times $x^{2(1-\frac{1}{\alpha})}$ then the probability of the increment being positive is negligibly small. When the distance is less than that then a threshold of $g(\kappa) = c \log x$, for a large c, will be sufficient. This is so since the standard deviation of $w_\theta(\vartheta)$ is only of the order of magnitude of the square root of the threshold and the expectation is negative. Consequently, the probability of being larger than the threshold is again negligible.

9.3.10   Checking Condition II*

Condition II* also uses Theorem A.4 and the method of proof that was used in order to establish the approximation of T by the sparse grid $\hat{T}$. Specifically, we eliminate the terms that are associated with distant ϑ using an exponential bound on the tail of the distribution of $w_\theta(\vartheta)$ as a function of ϑ. For ϑ in the vicinity of θ (but outside of the local region T_θ) we consider the intersection of the event $\{w_\theta(\vartheta) \geq \log(\epsilon p_\vartheta)\}$ and the event $\{\tilde{\ell}_\theta \in (-m, \delta]\}$. We condition on the σ-algebra that is generated by the source-specific demands that are associated with the interval θ. The conditional probability of the event associated with $w_\theta(\vartheta)$ is bounded by a sum of an indicator of a measurable event and the bound produced by the concentration inequality of Theorem A.4. A bound on the density of $\tilde{\ell}_\theta$ multiplies the outcome of the concentration inequality by a term that is equal to $x^{1-\frac{1}{\alpha}}$ times a universal constant.

The other term in the bound is the indicator that was left out. This indicator involves a statement regarding the conditional variance and the conditional expectation of $w_\theta(\vartheta)$, given σ-algebra of the source-specific demands. The statement regarding the conditional variance is dealt with directly using large deviation with respect to n. For the statement regarding the conditional expectation we apply Theorem 5.3 locally to $\tilde{\ell}_\theta$, jointly with the asymptotically independent conditional expectation of $w_\theta(\vartheta)$. Selection of a large enough τ will give a small sum of probabilities, even after multiplying each probability with $x^{1-\frac{1}{\alpha}}$. This completes the proof of Condition II* and the validation of Theorem 5.2.

9.4   Heavy tail distribution, long-range dependence, and self-similarity

The approximation of the probability dependent on the parameter of sparseness δ via the term:

$$\delta^{-2}[\mathrm{E}(\mathcal{M}/\mathcal{S})]^2 = [\mathrm{E}(\mathcal{M}/(\delta\mathcal{S}))]^2 .$$

It is time now to examine the limit, as $\delta \to 0$, of this term. With the resulting limit we will be able to produce an expression for the probability of an overflow as it was originally formulated.

The expectation of the ratio between the maximum and the sum is defined in (9.13). The maximization and summation are with respect to an exponentiated double-ended Gaussian process. The Gaussian process is given by $\sigma_\delta W_i - 0.5\sigma_\delta^2 |i|^\alpha$, defined for all negative and positive integers. The random part W_i is a centered Gaussian process with stationary increments that is uniquely characterized by the variance function $\mathrm{Var}(W_i) = |i|^\alpha$. The dependence of the exponentiated process on δ is via the multiplicative scaling parameter $\sigma_\delta = 2v[\mu(\alpha)]\delta^{\frac{\alpha}{2}}$.

The process $\sigma_\delta W_i$ is, again, centered with stationary increments. As such, it is fully characterized by the variance function, which is equal to

$$\mathrm{Var}(\sigma_\delta W_i) = \sigma_\delta^2 |i|^\alpha = 4v^2[\mu(\alpha)]^2\delta^\alpha |i|^\alpha = 2\left(\left[\sqrt{2}v\mu(\alpha)\right]^{\frac{2}{\alpha}}\delta|i|\right)^\alpha .$$

This variance structure is identical to the variance structure of the process: $\sqrt{2}W_{\zeta\delta i}$, for $\zeta = [\sqrt{2}v\mu(\alpha)]^{\frac{2}{\alpha}}$, that is defined over the refined grid with span $\zeta\delta$. The drift can also be reformulated over the given grid, since $0.5\sigma_\delta^2 |i|^\alpha = |\zeta\delta i|^\alpha$. Equality in distribution leads to the conclusion that:

$$\mathrm{E}(\mathcal{M}/(\delta\mathcal{S})) = \mathrm{E}\left[\frac{\max_{t \in T_\delta} e^{\sqrt{2}W_t - |t|^\alpha}}{\delta\sum_{t \in T_\delta} e^{\sqrt{2}W_t - |t|^\alpha}}\right],$$

for $T_\delta = \{\zeta\delta i : i \in \mathbb{Z}\}$. Like before, the process W is a centered Gaussian process with stationary increments that is characterized by the variance function $\mathrm{Var}(W_t) = |t|^\alpha$.

The given process W_t, considered over the entire real line, has a version with continuous paths. We use that version. When we let δ go to 0 the maximization of the exponentiated process converges to the maximum of the process $\exp\{\sqrt{2}W_t - |t|^\alpha\}$. The Riemann sum of the exponentiated process converges to the integral. In the next chapter we prove uniform integrability of the ratio, which leads to the conclusion that:

$$\lim_{\delta \to 0} E(\mathcal{M}/(\delta\mathcal{S})) = \zeta \cdot E\left[\frac{\max_{t \in \mathbb{R}} e^{\sqrt{2}W_t - |t|^\alpha}}{\int_{-\infty}^{\infty} e^{\sqrt{2}W_t - |t|^\alpha} dt}\right] = [\sqrt{2}v\mu(\alpha)]^{\frac{2}{\alpha}} \cdot \mathcal{H}_\alpha .$$

The constant $\mathcal{H}_\alpha$, that depends on the parameter of long-range dependence α, is the expectation of the ratio between the maximum and the integrated exponentiated continuous process. This constant, with a different representation, was discovered by Pickands as part of his work in developing the double-sum method for Gaussian processes. In the next chapter we present the original representation and discuss the issue of evaluating the constants. Here we use these constants in order to give the final representation of the probability of a buffer overflow:

$$P\left(\max_{\theta \in T}(Y_\theta - \mu|\theta|) \geq x\right) \approx m e^{-\frac{x^{2-\alpha}\mu(\alpha)}{(1-\alpha/2)}} x^{\frac{4}{\alpha}+\alpha-5} \frac{[\sqrt{2}v\mu(\alpha)]^{\frac{4}{\alpha}} v^2(\alpha/2)^{\alpha+\frac{1}{2}}}{\mu^{\alpha+1}(1-\alpha/2)^{\alpha-\frac{1}{2}}} \mathcal{H}_\alpha^2 ,$$

(9.14)

for $T = \{\theta = (t_1, t_2] : 0 \leq t_1 < t_2 \leq m, t_1, t_2 \in \mathbb{N}\}$.

The approximation of the probability of an overflow that is given in (9.14) is a function of the length of the time interval m, the threshold x, the asymptotic variance v^2, and μ, the rescaled difference between bandwidth and expected demand. What makes this expression more interesting for the next chapter is its dependence on the parameter α that characterizes long-range dependence.

Before leaving this example and moving to a more general discussion we may add the remark that, on the basis of the approximation (9.14), one can derive a Poisson approximation that is relevant when m is large. For example, in the Gaussian limit the proof can use a truncation argument to eliminate extremely long intervals, bounding the value of $|\theta|$ as a function of the threshold x. The Slepian inequality may then be used in order to bound the probability from above using a process that has larger variability, since far removed increments are independent in that other process. The same inequality can be used in order to bound the probability from below with a process that has lesser variability but a simpler covariance structure. Specifically, we consider the process where all increments with distance above a given threshold have the same low but fixed covariance. The Poisson approximation can be applied to the lower and the upper bound using the method that was described in Chapter 4.

A clump is associated with a continuous period of time where the system suffered from a buffer overflow. This puristic description needs modification since, especially at the edges, one may observe some variations below and above the threshold which should be included in the clump. With this correction one may identify that the number of clumps follows the Poisson distribution that was identified as a result of (9.14). Reversing Aldous' argument that was mentioned in Section 2.4 we may propose that the expected length of a clump can be approximated by the ratio between the expectation of the total number of intervals above the threshold, divided by the expected number of clumps. The expected number of clumps is given by (9.14) and the expectation of the total number is the sum of the marginal probabilities.

Let us leave the specific application and return to the general discussion. The concept of long-range dependence is more naturally formalized in the context of processes with stationary increments. If these increments are uncorrelated then the variance of an increment of a given length is equal to the sum of variances of sub-increments of length 1 each and thus is linear in the length of the interval. This linearity may not be exact when increments are correlated but may still be asymptotically so if the correlation is reduced fast enough as a function of the distance between the sub-increments. Otherwise, if the reduction is slow enough to produce an exponent other than 1 in the relation between the variance of an increment the length of the interval, we define the process to possess long-range dependence.

Long-range dependence is an asymptotic property and was formulated in the current setting in terms of the exponent α in the relation $\text{Var}(X_t) \sim v^2 t^\alpha$. An exponent in the range $1 < \alpha < 2$ corresponds to a positive correlation between increments. The extreme case is $\alpha = 2$ where increments are fully correlated. The range of the definition of the parameter α can be extended to the range $0 < \alpha < 1$, in which case it corresponds to processes that are more variable than processes of partial sums of uncorrelated random variables.

In the particular example that was analyzed in this chapter long-range dependence emerged as a result of the heavy tail distribution of duration times in either of the states of the demand on-off process. That may be the case, but it should still be remembered that conceptually, long-range dependence and heavy tail distribution are unrelated; one talks about a correlation structure in a stochastic process and the other about the marginal distribution of a random variable. For example, in the case of the process Y_t long-range dependence is present. However, if we consider the marginal distribution of the component Y_t then that distribution is a scaled sum of independent and bounded random variables, in the limit a Gaussian random variable, in both cases a variable with a light tail.

Another concept that is frequently associated with long-range dependence is the concept of self-similarity. Actually, in a way, self-similarity may be associated with stationarity. If stationarity holds then one may change the location of the origin without changing the distribution. The process that emerges from shifting the time by some fixed amount has the same distribution as the original process. Self-similarity is a similar phenomenon in the context of changing the timescale. A process is self-similar if after changing the timescale by a fixed amount, and at the same time multiplying the process by a constant that restores the variance of increments, the transformed process has the same distribution as the original process.

An example of a self-similar process is Brownian motion. Changing the timescale by c, and multiplying the process by $c^{-\frac{1}{2}}$ reproduces the original distribution. The resulting process has the same 0 expectation and the same covariance structure as the original process, albeit in the new timescale. The Gaussian processes W_t that are used to describe Pickands' constants are also self-similar. Changing the timescale by c, and multiplying the process by $c^{-\frac{\alpha}{2}}$ will reproduce the original variance for the increments. Moreover, since the processes are stationary, that will imply an identical covariance structure, hence equality in distribution. As it turns out, the only stationary Gaussian processes with continuous paths are those with variance proportional to t^α, for some $0 < \alpha \leq 2$. These are the processes that emerge as the limit of the local field in the Gaussian and asymptotic Gaussian cases. Previously we met the Brownian motion type of limit ($\alpha = 1$) and the smooth type of limit ($\alpha = 2$). Now we may add to the collection all the processes that are associated with the other values of α. These other members are called fractional Brownian motions.

The constant that emerges in the simple, one-dimensional, case as the limit of the local field is $\mathcal{H}_\alpha$. This constant captures the effect of local fluctuation on the probability of crossing a high threshold. In order to produce a usable formula for the approximation of these probabilities we need to be able to evaluate the numerical values of these constants. For many years this was an unobtainable goal. In the next chapter we describe how it can be done.

10

Computing Pickands' constants

10.1 Introduction

In the previous chapter we investigated a system that involved long-range dependence. A special type of constants, Pickands' constants $\mathcal{H}_\alpha$ emerged as part of the expression for the probability that the extreme of the field in question exceeds a high threshold. The constants were represented as the expectation of the ratio between the maximum and the sum (or integral), with both numerator and denominator produced by Gaussian likelihood ratios that are connected to the fractional Brownian motion. This chapter is devoted to the computation of these constants. Most of the theoretical derivation in this chapter is taken from [29].

But before we do that let us examine these constants in their historical context. Originally, the constants surfaced as a consequence of the double-sum method, the method for analyzing maxima in Gaussian fields that was invented by Pickands in the late 1960s. As a matter of fact, we have already met a representative of these constants when we introduced the double-sum method in order to investigate the probability of false detection in a scanning statistic that involves an indicator function serving as the kernel. Actually, we could have also used the same method in order to investigate the case where the kernel is a smooth function, where we applied a different tool, and obtain another member of the same family of constants.

The main characteristic of both examples in the current context is the behavior of the correlation function near the diagonal. On the diagonal $\vartheta = \theta$ the correlation is 1. An asymptotic expansion of the covariance for values of ϑ close to θ gave $\mathrm{Cov}(Z_\vartheta, Z_\theta) = 1 - c\|\vartheta - \theta\|^\alpha + o(\|\vartheta - \theta\|^\alpha)$, with α taking the value 2 in the smooth case and the value 1 when an indicator is involved.

Extremes in Random Fields: A Theory and its Applications, First Edition. Benjamin Yakir.

We would like to expand the approximation of the tail of a Gaussian field to all values of α, $0 < \alpha \leq 2$. It would be good enough for our goal to consider the case of a standardized Gaussian process, a one-dimensional field. Here the parameter θ is a number in an interval T. Let us assume the expectation of the process $\{Z_\theta : \theta \in T\}$ is identically 0, the variance is identically 1, and the covariance is strictly less than 1 away from the diagonal and has the expansion $\mathrm{Cov}(Z_\vartheta, Z_\theta) = 1 - |\vartheta - \theta|^\alpha + o(|\vartheta - \theta|^\alpha)$ in the vicinity of the diagonal. First we want to apply the double-sum method in order to obtain an approximation of the probability that the maximum of the field is larger than z, for an increasing z. Subsequently, we would like to repeat the same analysis using our tool and compare the results. Since we have already the experience of applying both tools in more complicated examples, we will only sketch out the argument.

10.1.1   The double-sum method

In the double-sum method we start by splitting the interval of length $|T|$ in to $|T|z^{\frac{2}{\alpha}}/\tau$ sub-intervals of length $\tau z^{-\frac{2}{\alpha}}$ each. We denote each sub-interval T_θ, after the index value of the left-endpoint of the interval, and consider the probability of the event $\{\max_{\vartheta \in T_\theta} Z_\vartheta \geq z\}$. This probability is computed by conditioning on the value of Z_θ:

$$P\left(\max_{\vartheta \in T_\theta} Z_\vartheta \geq z\right) = \frac{1}{\sqrt{2\pi}} \int e^{-\frac{1}{2}y^2} P\left(\max_{\vartheta \in T_\theta} Z_\vartheta \geq z \mid Z_\theta = y\right) dy .$$

After a change in the integration variable $x = -z(y - z)$ we get:

$$= \frac{e^{-\frac{1}{2}z^2}}{\sqrt{2\pi}z} \int e^{x - \frac{1}{2}(\frac{x}{z})^2} P\left(\max_{\vartheta \in T_\theta} z(Z_\vartheta - Z_\theta) \geq x \mid Z_\theta = z - x/z\right) dx .$$

Next we investigate the conditional distribution of the local field $z(Z_\vartheta - Z_\theta)$. The conditional expectation is:

$$E(z(Z_\vartheta - Z_\theta)|Z_\theta = z - x/z) = -zZ_\theta(1 - \mathrm{Cov}(Z_\vartheta, Z_\theta)) \approx -(z^2 - x)|\vartheta - \theta|^\alpha$$

and the conditional variance of an increment is:

$$\mathrm{Var}(z(Z_\vartheta - Z_\eta)|Z_\theta) = z^2\{2[1 - \mathrm{Cov}(Z_\vartheta, Z_\eta)] - [\mathrm{Cov}(Z_\vartheta, Z_\theta) - \mathrm{Cov}(Z_\eta, Z_\theta)]^2\}$$

$$= 2z^2|\vartheta - \eta|^\alpha + o(|\vartheta - \eta|^\alpha) .$$

Denoting $t = (\vartheta - \theta)z^{2/\alpha}$, that takes values in the range $[0, \tau]$, we get that the limit process can be represented as $\sqrt{2}W_t - t^\alpha$, where W_t has stationary increments and, thereby, the covariance structure of the fractional Brownian motion:

$$\mathrm{Cov}(W_s, W_t) = \frac{1}{2}(|s|^\alpha + |t|^\alpha - |t - s|^\alpha) .$$

When we pass to the limit the probability of crossing the threshold in the interval T_θ becomes:

$$\sqrt{2\pi}\, z^{-\frac{2}{\alpha}+1} e^{\frac{1}{2}z^2} \mathrm{P}\left(\max_{\vartheta \in T_\theta} Z_\vartheta \geq z\right) \approx \int e^x \mathrm{P}\left(\max_{0 \leq t \leq \tau} [\sqrt{2}W_t - t^\alpha] \geq x\right) dx = \mathcal{H}_\alpha(\tau),$$

for

$$\mathcal{H}(\tau) = \mathrm{E}\left(\max_{0 \leq t \leq \tau} e^{\sqrt{2}W_t - t^\alpha}\right).$$

In order to produce the upper bound in the double-sum we need to sum the approximations over all the sub-intervals and pass to a limit in τ. The sum is asymptotically bounded by:

$$\mathrm{P}\left(\max_{\vartheta \in T} Z_\vartheta \geq z\right) \lesssim |T|(2\pi)^{-\frac{1}{2}} z^{\frac{2}{\alpha}-1} e^{-\frac{1}{2}z^2} [\mathcal{H}_\alpha(\tau)/\tau].$$

As we let τ grow to infinity we obtain the originally derived representation of Pickands' constant:

$$\mathcal{H}_\alpha = \lim_{\tau \to \infty} \frac{1}{\tau} \mathcal{H}_\alpha(\tau) = \lim_{\tau \to \infty} \frac{1}{\tau} \mathrm{E}\left(\max_{0 \leq t \leq \tau} e^{\sqrt{2}W_t - t^\alpha}\right). \tag{10.1}$$

A full proof, which is based on the double-sum method, will have to construct a lower bound which is based on the double-sum and show that the double-sum is negligible when $\tau \to \infty$. We did not provide a proof for that fact in our first application of the double-sum. We will not provide a proof now. Instead, we will start over with an attempt to approximate the probability that the maximum of the process over the interval exceeds z using our approach.

10.1.2 The method based on the likelihood ratio identity

The first step is to approximate the original continuous interval T by a dense collection of points $\hat{T} \subset T$. We apply the inequality:

$$\mathrm{P}\left(\sup_{\theta \in T} Z_\theta \geq z\right) \leq \mathrm{P}\left(\max_{\theta \in \hat{T}} Z_\theta \geq z - \frac{\epsilon}{z}\right) + \mathrm{P}\left(\max_{\theta \in \hat{T}} Z_\theta \leq z - \frac{\epsilon}{z}, \sup_{\vartheta \in T} Z_\vartheta \geq z\right)$$

in order to prove the accuracy of the approximation via a bound produced to the very last probability. If we denote the points in the collection $\hat{T}$ by θ and define this time a local region T_θ to involve the points in the interval $[\theta, \theta + \delta z^{-\frac{2}{\alpha}}]$ we get that the last probability is bounded by:

$$\mathrm{P}\left(Z_\theta \leq z - \frac{\epsilon}{z}, \sup_{\vartheta \in T_\theta} Z_\vartheta \geq z\right)$$

$$= e^{-\frac{1}{2}z^2} \frac{1}{\sqrt{2\pi}\, z} \int_{-\infty}^{-\epsilon} e^{-y - \frac{y^2}{2z^2}} \mathrm{P}\left(\sup_{\vartheta \in T_\theta} z(Z_\vartheta - z) \geq 0 \,\middle|\, z(Z_\theta - z) = y\right) dy.$$

Again, the local field $z(Z_\vartheta - z)$, conditional on $z(Z_\theta - z) = y$, converges to a limit Gaussian field with the covariance structure of the fractional Brownian motion. The conditional expectation is:

$$E(z(Z_\vartheta - z)|z(Z_\theta - z) = y) = y - z^2[1 - \text{Cov}(Z_\vartheta, Z_\theta)]\left[1 + y/z^2\right] ,$$

which is asymptotically bounded from above over the local interval by $-x(y, z) =$
$y - \delta^\alpha[1 + y/z^2]$.

$$P\left(\sup_{\vartheta \in T_\theta} z(Z_\vartheta - z) \geq 0 | z(Z_\theta - z)\right) \lesssim P\left(\sup_{0 \leq t \leq \delta} W_t \geq x(y, z)\right)$$

$$\leq B_\alpha e^{-\frac{1}{2}[x(y,z)]^2/[C_\alpha^2 \delta^\alpha]} ,$$

which results from the application of Fernique's inequality to the fractional Brownian motion W_t.

Taking into account the fact that the total number of sub-intervals is $|T| \cdot \delta^{-1} z^{\frac{2}{\alpha}}$ we get the bound:

$$e^{\frac{1}{2}z^2} z^{-\frac{2}{\alpha}+1} P\left(\max_{\theta \in \hat{T}} Z_\theta \leq z - \frac{\epsilon}{z}, \sup_{\vartheta \in T} Z_\vartheta \geq z\right)$$

$$\lesssim |T| \cdot \frac{B_\alpha}{\delta\sqrt{2\pi}} \int_\epsilon^\infty e^{y - \frac{y^2}{2z^2}} e^{-\frac{1}{2}[x(y,z)]^2/[C_\alpha^2 \delta^\alpha]} dy$$

$$\lesssim |T| \cdot \frac{B_\alpha C_\alpha}{\delta^{1-\frac{\alpha}{2}}} e^{(d + \frac{C_\alpha^2}{2})\delta^\alpha} \left[1 - \Phi\left(\frac{\epsilon - (d + C_\alpha^2)\delta^\alpha}{C_\alpha \delta^{\frac{\alpha}{2}}}\right)\right] .$$

For a fixed $\epsilon > 0$ this bound converges to 0 with δ.

We are in a position to apply the likelihood ratio identity. Calling the newly created discrete subset of parameters T and using the fact that $zZ_\theta - z^2/2$ is a log-likelihood ratio for the entire process we obtain the representation:

$$P\left(\max_{\theta \in T} Z_\theta \geq z\right) = e^{-\frac{1}{2}z^2} \sum_{\theta \in T} E_\theta\left(\frac{M_\theta}{S_\theta} e^{-[\tilde{\ell}_\theta + m_\theta]}; \tilde{\ell}_\theta + m_\theta \geq 0\right) .$$

The terms M_θ and S_θ are functions of the local field $\{z(Z_\vartheta - Z_\theta)\}$, defined over the discrete grid T. The global term is $\tilde{\ell}_\theta = z(Z_\theta - z)$, a centered Gaussian random variable under the tilted distribution with z^2 as the variance.

The local field converges, as $z \to \infty$, to the double-ended fractional Brownian motion with negative drift $\sqrt{2}W_t - |t|^\alpha$, for $t \in \{ \ldots, -2\delta, -\delta, 0, \delta, 2\delta, \ldots \}$. As a result of the application of the localization theorem we get that the contribution of the local fluctuation to the limit is:

$$E(\mathcal{M}_\delta \mathcal{S}_\delta) = E\left(\frac{\max_{i \in \mathbb{Z}} e^{\sqrt{2}W_{i\delta} - |i\delta|^\alpha}}{\sum_{i \in \mathbb{Z}} e^{\sqrt{2}W_{i\delta} - |i\delta|^\alpha}}\right) .$$

The contribution of the global term is its density evaluated at the origin, namely $1/[\sqrt{2\pi}z]$. When we apply the integration step, which corresponds to a summation over T we obtain an approximation for the probability of extreme values:

$$P\left(\max_{\vartheta\in T}Z_\vartheta\geq z\right)\approx|T|(2\pi)^{-\frac{1}{2}}z^{\frac{2}{\alpha}-1}e^{-\frac{1}{2}z^2}\delta^{-1}E(\mathcal{M}_\delta\mathcal{S}_\delta)\ .$$

When $\delta\to 0$ we obtain by the continuity of the paths of the processes that $\max_{i\in\mathbb{Z}}\exp\{\sqrt{2}W_{i\delta}-|i\delta|^\alpha\}$ converges to $\max_{t\in\mathbb{R}}\exp\{\sqrt{2}W_t-|t|^\alpha\}$. Likewise, we get the convergence of the Riemann sum $\delta\sum_{i\in\mathbb{Z}}\exp\{\sqrt{2}W_{i\delta}-|i\delta|^\alpha\}$ to the integral $\int\exp\{\sqrt{2}W_t-|t|^\alpha\}dt$. Comparing the terms obtained by the approximation produced by the double-sum method with the terms produced by the current method, and depending on showing uniform integrability of the sequence of the ratio between the maxima and the sum times δ, we obtain an alternative representation for Pickands' constants:

$$\mathcal{H}_\alpha=E\left(\frac{\max_{t\in\mathbb{R}}e^{\sqrt{2}W_t-|t|^\alpha}}{\int e^{\sqrt{2}W_t-|t|^\alpha}dt}\right)\ . \tag{10.2}$$

Establishing uniform integrability will be part of the mathematical development of in this chapter.

10.1.3 Pickands' constants

The values of Pickands' constants were known only in the two cases that were discussed previously. In the case $\alpha=2$ that involves a smooth process one gets that $\mathcal{H}_2=1/\sqrt{\pi}$. In the case $\alpha=1$ of the regular Brownian motion the value is $\mathcal{H}_1=1$. Another fact that was proved in the past was that the constants converge to 0 when $\alpha\to 0$. Based on these facts it was conjectured at some points that the constants may correspond to the function $f(\alpha)=1/\Gamma(1/\alpha)$, which matches the know values of the constants at $\alpha=0$, 1, and 2 and, overall, looks like a nice function. Theoretical investigations were able to produce upper and lower limits for the values of the constants, but were not refined enough to discredit the conjecture.

Simulation studies were also tried. It should be noted that both representations (10.1) and (10.2) may not be used directly for simulations. There exist efficient algorithms for generating sample paths of a fractional Brownian motion. However, all methods are limited to the production of the values of the process over a discrete and finite grid. Both formulae are given in terms of a continuous and infinite process. Consequently, any method that relies on simulations computes, in actuality, functionals of a finite and discrete approximation of the process, and not of the process directly. In other words, a numerical evaluation via simulations calls for a truncation of the parameter space and its approximation by a discrete subset.

In the case of the representation (10.1) this may correspond to setting the value of τ to be some finite value and substituting the continuous process by a discrete approximation producing $\tau^{-1}\max_{0 \le i \le \tau/\delta} \exp\{\sqrt{2}W_{i\delta} - |i\delta|^\alpha\}$. The expectation of that term may be approximated by averaging over simulated sample paths. In the case of the representation (10.2) one may set again a finite value for τ and consider the ratio $\max_{0 \le i \le \tau/\delta} \exp\{\sqrt{2}W_{i\delta} - |i\delta|^\alpha\}/\delta \sum_{i=0}^{\tau/\delta} \exp\{\sqrt{2}W_{i\delta} - |i\delta|^\alpha\}$. The average over the simulated sample paths of this ratio can serve as an approximation of the expectation.

The accuracy of simulations should be addressed as a problem in statistical inference. In either approach for the simulation of the constants one should worry about the bias, i.e., the discrepancy between the expectation of the simulated value and the constant, and the variance that is involved in the simulation. The main problem in the simulation that is based on the representation (10.1) is the variance. Typically, the maximum of the entire simulated process will occur at values of i close to the origin and typically the value of the maximum will not be too large. However, occasionally the maximum takes place at remote locations and the maximal value is large. This large value is exaggerated further by taking an exponent. The consequence is a variance that grows rapidly with the increase of τ.

Luckily enough, the simulation approach that is based on representation (10.2) does not suffer from the same weakness. When a large value occurs in the maximum it occurs also in the sum. As a result, the increase in the variance that results from increasing τ is not as acute, which results in a trustworthy simulation procedure.

In this chapter we will construct bounds on the errors that may be produced by simulations that use (10.2). These bounds will be developed Section 10.3 that deals with the statistical properties of the simulations. Before that, in the next section, we will produce several representations of Pickands' constants and related constants. In section 10.4 we will discuss a potential approach for extending approximations to discrete processes in the presence of long-range dependence.

10.2   Representations of constants

This section connects directly the two representations of Pickands' constants that are given in (10.1) and (10.2). Subsequently, we give other representations of the constants and of related constants. The basis for these different representations is a lemma for a change-of-measure. For this lemma we denote $Z_t = \sqrt{2}W_t - |t|^\alpha$, where W_t is the fractional Brownian motion evaluated at t, and denote the entire process by Z. Define the time-shift operator θ_t via $(\theta_t Z)_s = Z_{s-t}$. Then:

Lemma 10.1   *Fix $t \in \mathbb{R}$, and set $Z^{(t)} = \{\sqrt{2}W_s - |s - t|^\alpha : s \in \mathbb{R}\}$. For an arbitrary functional F on $\mathbb{R}^\mathbb{R}$, we have:*

$$E(e^{Z_t} F(Z)) = EF(|t|^\alpha + Z^{(t)}) .$$

If F is translation-invariant (invariant under addition of a constant function) then:

$$E(e^{Z_t} F(Z)) = EF(\theta_t Z).$$

Proof. Select an integer k and $s_1 < s_2 < \cdots < s_k$, and note that for any $\xi_1, \ldots, \xi_k \in \mathbb{R}$,

$$\log \mathrm{E}\left[e^{Z_t} \exp\left(\sum_{i=1}^{k} \xi_i Z_{s_i} \right) \right]$$

$$= -|t|^\alpha - \sum_{i=1}^{k} \beta_i |s_i|^\alpha + \mathrm{Var}\left(W_t + \sum_{i=1}^{k} \xi_i W_{s_i} \right)$$

$$= \sum_{i=1}^{k} 2\xi_i \mathrm{Cov}(W_t, W_{s_i}) - \sum_{i=1}^{k} \xi_i |s_i|^\alpha + \mathrm{Var}\left(\sum_{i=1}^{k} \xi_i W_{s_i} \right)$$

$$= \sum_{i=1}^{k} \xi_i (|t|^\alpha - |s_i - t|^\alpha) + \mathrm{Var}\left(\sum_{i=1}^{k} \xi_i W_{s_i} \right)$$

$$= \sum_{i=1}^{k} \xi_i |t|^\alpha + \mathrm{E}\left(\sum_{i=1}^{k} \xi_i Z_{s_i}^{(t)} \right) + \frac{1}{2}\mathrm{Var}\left(\sum_{i=1}^{k} \xi_i Z_{s_i}^{(t)} \right) .$$

The multivariate moment generating function determines uniquely the multivariate distribution. In the current case, we obtained the multivariate moment generating function of a Gaussian process. The distribution of the entire process is generated by finite-dimensional multivariate distributions. Consequently, the distribution of the entire process is set, and therefore the distribution of the image of a functional. This proves the first part of the lemma.

If the functional F is translation-invariant then we can add and delete terms that depend on a single time-point t and conclude as required that:

$$\mathrm{E}(e^{Z_t} F(Z)) = \mathrm{E}F(|t|^\alpha + Z^{(t)}) = \mathrm{E}F(Z^{(t)} - \sqrt{2}W_t) = \mathrm{E}F(\theta_t Z) . \qquad \square$$

We will use this lemma in order to produce alternative representations of Pickands' constants and constants of the same type. The relation to the lemma is the fact that the functional that takes the ratio between the maximum of the exponentiated process and the sum (or integral) of the same exponentiated process is translation-invariant.

In particular, we can get as a corollary of this lemma a direct link between the two representations of the Pickands constant:

$$\frac{1}{\tau}\mathrm{E}\left(\sup_{0 \leq t \leq \tau} e^{Z_t} \right) = \int_0^1 \mathrm{E}\left(\frac{\sup_{-u\tau \leq s \leq (1-u)\tau} e^{Z_s}}{\int_{-u\tau}^{(1-u)\tau} e^{Z_s} ds} \right) du , \qquad (10.3)$$

for any $\tau > 0$.

Indeed, when we apply Lemma 10.1 to the translation-invariant functional

$$F(z) = \frac{\sup_{t \in [0,\tau]} e^{z_t}}{\int_0^\tau e^{z_u} du}$$

we get that:

$$\frac{1}{\tau}E\left(\sup_{0 \le t \le \tau} e^{Z_t}\right) = \frac{1}{\tau}\int_0^\tau E\left(e^{Z_t} \times \frac{\sup_{0 \le s \le \tau} e^{Z_s}}{\int_0^\tau e^{Z_s}ds}\right)dt$$

$$= \frac{1}{\tau}\int_0^\tau E\left(\frac{\sup_{-t \le s \le \tau-t} e^{Z_s}}{\int_{-t}^{\tau-t} e^{Z_s}ds}\right)dt \ .$$

Relation (10.3) follows from a change of variable $t = u\tau$.

When we let $\tau \to \infty$ we get that the left-hand side of (10.3) converges to definition (10.1) of $\mathcal{H}_\alpha$. On the other hand, for each fixed u the integrand on the right-hand side converges to definition (10.2) of the same constants. In order to establish rigorously the convergence of the entire integral to expression (10.2) one needs to validate uniform integrability. This validation is a corollary of the truncation argument that we give in the next section.

We can easily obtain other representations of the constants. For example if we use the functional:

$$F_\delta(z) = \frac{\sup_{t \in [0,\tau]} e^{z_t}}{\delta \sum_{i=0}^{\tau/\delta} e^{z_{i\delta}}}$$

and follow the same steps that led from the representation of the functional F to the establishment of the equivalency between (10.1) and (10.2) then we will obtain the fact that:

$$\mathcal{H}_\alpha = E\left(\frac{\max_{t \in \mathbb{R}} e^{\sqrt{2}W_t - |t|^\alpha}}{\delta \sum_{i \in \mathbb{Z}} e^{\sqrt{2}W_{i\delta} - |i\delta|^\alpha}}\right), \tag{10.4}$$

for any $\delta > 0$.

We can explore relations between quantities going in the other direction. If we consider the constant that emerged as a result of applying the likelihood ratio technique to the process over the approximating discrete grid:

$$\mathcal{H}_\alpha^\delta = E\left(\frac{\max_{i \in \mathbb{Z}} e^{\sqrt{2}W_{i\delta} - |i\delta|^\alpha}}{\delta \sum_{i \in \mathbb{Z}} e^{\sqrt{2}W_{i\delta} - |i\delta|^\alpha}}\right) = \lim_{\tau \to \infty} \frac{1}{\tau}\sum_{u=0}^{\tau/\delta} E\left(\frac{\max_{-u \le i \le \tau/\delta - u} e^{Z_{i\delta}}}{\delta \sum_{i=-u}^{\tau/\delta - u\delta} e^{Z_{i\delta}}}\right)$$

We can conclude that

$$\mathcal{H}_\alpha^\delta = \lim_{\tau \to \infty} \frac{1}{\tau}E\left(\max_{0 \le i \le \tau/\delta} e^{\sqrt{2}W_{i\delta} - (i\delta)^\alpha}\right). \tag{10.5}$$

The right-hand side of (10.5) is the Pickands constant when the process is restricted to a discrete grid. Off course, one may use the functional

$$F_\delta(z) = \frac{\max_{0 \le i \le \tau/\delta} e^{z_t}}{\int_0^\tau e^{z_t}dt}$$

and obtain a different representation of the same constant:

$$\mathcal{H}_\alpha^\delta = \mathrm{E}\left(\frac{\max_{i\in\mathbb{Z}}e^{\sqrt{2}W_{i\delta}-|i\delta|^\alpha}}{\int e^{\sqrt{2}W_t-|t|^\alpha}\,dt}\right). \tag{10.6}$$

10.3   Analysis of statistical error

The main goal is to evaluate the constants $\mathcal{H}_\alpha$ via simulations. An iteration of the simulation produces a path of the fractional Brownian motion with negative drift $Z_{i\delta} = W_{i\delta} - |i\delta|^\alpha$ evaluated on a grid with span δ and over the interval $[-\tau, \tau]$. This path can be used in order to compute the statistic that divides the maximal exponentiated process by δ times the sum of the exponentiated process. This statistic is an unbiased estimate of the constant:

$$\mathcal{H}_\alpha^\delta(\tau) = \mathrm{E}\left(\frac{\max_{|i|\leq\tau/\delta}e^{\sqrt{2}W_{i\delta}-|i\delta|^\alpha}}{\delta\sum_{i=-\tau/\delta}^{\tau/\delta}e^{\sqrt{2}W_{i\delta}-|i\delta|^\alpha}}\right). \tag{10.7}$$

The variance of this statistic, which has values restricted to the interval $[0, 1/\delta]$, is bounded by $(2\delta)^{-2}$. In actuality, the variance is much less and can be estimated as part of the simulation. The standard deviation in estimating (10.7) decreases in proportion to the reciprocal of the square root of the number of iterations of the simulation.

The main concern in planning the simulations is the discrepancy between $\mathcal{H}_\alpha^\delta(\tau)$ that is evaluated as a result of the simulations and the actual target $\mathcal{H}_\alpha$. This discrepancy is a function of the truncation parameter τ and the span of the grid δ. This section is devoted to the investigation of this discrepancy. The aim is to set bounds on the error of estimating the Pickands constant by a quantity that can be computed via simulation.

As part of the planning we select δ and use the equivalent definition of Pickands' constants that is given in (10.4). First we express in terms of τ and δ the discrepancy between the constant $\mathcal{H}_\alpha$ and a truncated version of this constant $\mathcal{H}_\alpha(\tau)$, in which maximization of the continuous exponentiated process and the Riemann sum of the discrete process are restricted to the interval $[-\tau, \tau]$. For the truncated constant $\mathcal{H}_\alpha(\tau)$ we have the inequality $\mathcal{H}_\alpha^\delta(\tau) \leq \mathcal{H}_\alpha(\tau)$, which follows from the fact that in the smaller constant the maximization over the entire interval is replaced by a maximization over a discrete sub-collection of points from the interval. In order to complete the analysis of the error an upper bound on $\mathcal{H}_\alpha(\tau)$ in terms of $\mathcal{H}_\alpha^\delta(\tau)$ is developed. This upper bound also depends on τ and δ. Finally, we select τ and δ on the basis of computational resources and technical limitations to obtain the smallest error terms.

The analysis of the error terms relies on a bound for the tail distribution of the maximum of a Gaussian process. In the current analysis we use Borell's inequality. The inequality is applied to a centered process. The parameters that determine the inequality are the maximal variance of the Gaussian process and the expectation of the maximum of the process. The latter expectation can be obtained via a chaining argument and entropy considerations.

Consider the process $Z_t = \sqrt{2} W_t - |t|^\alpha$ and the discrete approximation of this process $Z_t^\delta = Z_{\delta \lfloor t/\delta \rfloor}$, for positive t, and a similar definition for negative t. The centered version of the difference $Z_t - Z_t^\delta$ is $\sqrt{2}(W_t - W_t^\delta)$, where W_t^δ is the discrete approximation of W_t.

Consider a fixed interval $[t_0, t_1]$ and write:

$$\Delta(t_0, t_1) = \sup_{t_0 \leq t \leq t_1} (Z_t - Z_t^\delta) , \quad \Gamma(t_0, t_1) = \sqrt{2} \sup_{t_0 \leq t \leq t_1} (W_t - W_t^\delta) .$$

Observe that

$$\frac{\max_{t_0 \leq t \leq t_1} e^{\sqrt{2} W_t - |t|^\alpha}}{\delta \sum_{i=t_0/\delta}^{t_1/\delta} e^{\sqrt{2} W_{i\delta} - |i\delta|^\alpha}} = \frac{\mathcal{M}(t_0, t_1)}{\hat{\mathcal{S}}(t_0, t_1)} \leq \frac{1}{\delta} e^{\Delta(t_0, t_1)} . \qquad (10.8)$$

Therefore, given any event A, we have for $\log x > \mathrm{E}\Delta(t_0, t_1)$,

$$\mathrm{E}\left(\frac{\mathcal{M}(t_0, t_1)}{\hat{\mathcal{S}}(t_0, t_1)}; A\right)$$

$$\leq \mathrm{E}\left(\frac{\mathcal{M}(t_0, t_1)}{\hat{\mathcal{S}}(t_0, t_1)}; \frac{\mathcal{M}(t_0, t_1)}{\hat{\mathcal{S}}(t_0, t_1)} > \frac{x}{\delta}\right) + \frac{x}{\delta} \mathrm{P}(A)$$

$$= \frac{1}{\delta} \int_x^\infty \mathrm{P}\left(\frac{\mathcal{M}(t_0, t_1)}{\hat{\mathcal{S}}(t_0, t_1)} > \frac{y}{\delta}\right) dy + \frac{x}{\delta} \mathrm{P}\left(\frac{\mathcal{M}(t_0, t_1)}{\hat{\mathcal{S}}(t_0, t_1)} > \frac{x}{\delta}\right) + \frac{x}{\delta} \mathrm{P}(A)$$

$$\leq \frac{1}{\delta} \int_x^\infty \mathrm{P}(\Delta(t_0, t_1) \geq \log y) dy + \frac{x}{\delta} \mathrm{P}(\Delta(t_0, t_1) \geq \log x) + \frac{x}{\delta} \mathrm{P}(A) .$$

We are interested on a bound for this term.

Set

$$\kappa(t_1, t_2) = \sup_{t_1 \leq t \leq t_2} \{|t|^\alpha - (\delta \lfloor |t|/\delta \rfloor)^\alpha\} . \qquad (10.9)$$

Use the fact that $\Delta(t_0, t_1) \leq \Gamma(t_0, t_1) + \kappa(t_0, t_1)$ and the fact that the variance of $\Gamma(t_0, t_1)$ is bounded by $2\delta^\alpha$ to get by Borell's inequality that:

$$\mathrm{P}(\Delta(t_0, t_1) \geq \log y) \leq \exp\{-(\log y - \kappa(t_0, t_1) - \mathrm{E}\Gamma(t_0, t_1))^2 / [4\delta^\alpha]\} = G(y) . \qquad (10.10)$$

The only term that depends on the location is $\kappa(t_1, t_2)$. The distribution of $\Gamma(t_0, t_1)$ is a function of the length of the interval $t_1 - t_0$ but not of the specific location of the interval. The expectation $\mathrm{E}\Gamma(t_1, t_2) = \mathcal{E}(t_1 - t_0)$ can be bounded by a chaining argument that produces the upper bound:

$$\mathcal{E}(t_1 - t_0) = \frac{\sqrt{2\pi}}{\sqrt{\log(2)}} \sum_{j=2}^\infty 2^{3/2} (2\delta^{\frac{\alpha}{2}})^{j-1} \{\log(2|t_1 - t_0|^2 [2^{(\frac{2}{\alpha}) - \frac{1}{2}} \delta]^{-2j})\}^{\frac{1}{2}} . \qquad (10.11)$$

This upper bound is readily evaluated numerically.

To summarize, we have for each interval $[t_0, t_1]$, for any event A, and for all x such that $\log x > \mathcal{E}(t_1 - t_2) + \kappa(t_1, t_2)$ the inequality:

$$E\left(\frac{\mathcal{M}(t_0, t_1)}{\hat{\mathcal{S}}(t_0, t_1)}; A\right) \leq \frac{1}{\delta} \int_x^\infty G(y)dy + \frac{x}{\delta}G(x) + \frac{x}{\delta}P(A), \qquad (10.12)$$

where $G(y)$ is defined in (10.10), $\kappa(t_1, t_2)$ is defined in (10.9), and $\mathcal{E}(t_1 - t_2)$ is defined in (10.11).

We produce an error analysis of the truncation. Set $\mathcal{M}_0 = \mathcal{M}(-\tau, \tau)$ and $\hat{\mathcal{S}}_0 = \hat{\mathcal{S}}(-\tau, \tau)$. Fix $\gamma > 1$ and let $\mathcal{M}_j = \mathcal{M}(\tau\gamma^{j-1}, \tau\gamma^j)$, $\hat{\mathcal{S}}_j = \hat{\mathcal{S}}(\tau\gamma^{j-1}, \tau\gamma^j)$, for $j = 1, 2, \ldots$. Similar random variables can be defined for negative j.

Start with an upper bound on the difference between $E(\mathcal{M}/\hat{\mathcal{S}})$ and $E(\mathcal{M}_0/\hat{\mathcal{S}}_0)$. Use the fact that $\mathcal{S} \geq \mathcal{S}_j$ for any $j \in \mathbb{Z}$ to obtain:

$$E\left(\frac{\mathcal{M}}{\hat{\mathcal{S}}}\right) = E\left(\frac{\mathcal{M}_0}{\hat{\mathcal{S}}}; \mathcal{M} = \mathcal{M}_0\right) + \sum_{j \neq 0} E\left(\frac{\mathcal{M}_j}{\hat{\mathcal{S}}}; \mathcal{M} = \mathcal{M}_j\right)$$

$$\leq E\left(\frac{\mathcal{M}_0}{\hat{\mathcal{S}}_0}\right) + \sum_{j \neq 0} E\left(\frac{\mathcal{M}_j}{\hat{\mathcal{S}}_j}; \mathcal{M}_j > 1\right)$$

$$\leq E\left(\frac{\mathcal{M}_0}{\hat{\mathcal{S}}_0}\right) + 2\sum_{j=1}^\infty E\left(\frac{\mathcal{M}_j}{\hat{\mathcal{S}}_j}; \max_{\tau\gamma^{j-1} \leq t \leq \tau\gamma^j} \sqrt{2}W_t > \tau^\alpha \gamma^{(j-1)\alpha}\right).$$

$$(10.13)$$

In order to be able to implement (10.12) we need an upper bound on the probability of the event $A = \{\max_{\tau\gamma^{j-1} \leq t \leq \tau\gamma^j} \sqrt{2}W_t > \tau^\alpha\gamma^{(j-1)\alpha}\}$. By the scaling property of the fractional Brownian motion we get that the probability of this event is the same as the probability of the event $\{\max_{1 \leq t \leq \gamma} \sqrt{2}W_t > \tau^{\frac{\alpha}{2}}\gamma^{(j-1)\frac{\alpha}{2}}\}$. Again, we may apply Borell's inequality, exploiting the fact (see Theorem 2.8 in [30]) that $E(\max_{1 \leq t \leq \gamma} \sqrt{2}W_t) \leq \gamma^{\frac{\alpha}{2}}$ to get:

$$P(A) = P\left(\max_{\tau\gamma^{j-1} \leq t \leq \tau\gamma^j} \sqrt{2}W_t > \tau^\alpha\gamma^{(j-1)\alpha}\right) \leq \exp\{-(\tau^{\frac{\alpha}{2}}\gamma^{(j-1)\frac{\alpha}{2}} - \gamma^{\frac{\alpha}{2}})^2/(4\gamma^\alpha)\}.$$

$$(10.14)$$

For each $j \neq 0$ we may use (10.14) to bound the probability of the event. With this bound we may select $x = x_j$ that minimizes the bound in (10.12). The resulting bounds should be summed over all j.

The application of (10.12) does not vary by much the rate of convergence to zero that is given in (10.14). Even after the summation we get a rate of convergence to zero that is asymptotic to $\exp\{-c\tau^\alpha\}$ and is faster than any polynomial. This is in contrast with an attempt to use the original formulation of the Pickands constants for simulation. For the latter it can be shown in the case of the standard Brownian motion that the rate of convergence of the error produced by truncation is $\tau^{-\frac{1}{2}}$, which is much slower than the rate obtained by the current method.

The lower bound on the difference between the constant and the truncated version of the constant may be obtained in a similar method. This time we use:

$$
\mathrm{E}\left(\frac{\mathcal{M}}{\hat{\mathcal{S}}}\right) \geq \mathrm{E}\left(\frac{\mathcal{M}_0}{\hat{\mathcal{S}}_0} \cdot \frac{\hat{\mathcal{S}}_0}{\hat{\mathcal{S}}_0 + \sum_{j\neq 0}\hat{\mathcal{S}}_j}; \epsilon\hat{\mathcal{S}}_0 \geq \sum_{j\neq 0}\hat{\mathcal{S}}_j\right)
$$

$$
\geq \frac{1}{1+\epsilon}\mathrm{E}\left(\frac{\mathcal{M}_0}{\hat{\mathcal{S}}_0}; \epsilon\hat{\mathcal{S}}_0 \geq \sum_{j\neq 0}\hat{\mathcal{S}}_j\right)
$$

$$
= \frac{1}{1+\epsilon}\mathrm{E}\left(\frac{\mathcal{M}_0}{\hat{\mathcal{S}}_0}\right) - \frac{1}{1+\epsilon}\mathrm{E}\left(\frac{\mathcal{M}_0}{\hat{\mathcal{S}}_0}; \epsilon\hat{\mathcal{S}}_0 < \sum_{j\neq 0}\hat{\mathcal{S}}_j\right). \qquad (10.15)
$$

We seek to apply (10.12) once more. This time we set $A = \{\epsilon\hat{\mathcal{S}}_0 < \sum_{j\neq 0}\hat{\mathcal{S}}_j\}$. Since $\hat{\mathcal{S}}_0 \geq \delta$ we obtain:

$$
\mathrm{P}(A) \leq \mathrm{P}\left(\sum_{j\neq 0}\hat{\mathcal{S}}_j > \epsilon\delta\right) \leq 2\sum_{j=1}^{\infty}\mathrm{P}(\hat{\mathcal{S}}_j > \epsilon\delta q(1+q)^{-j})
$$

for some $q > 0$.

The number of terms that are involved in the summation that produces $\hat{\mathcal{S}}_j$ is $\tau\gamma^{j-1}(\gamma-1)/\delta$. The largest among these elements is bounded by:

$$
\exp\left\{\max_{\tau\gamma^{j-1}\leq t\leq\tau\gamma^j} \sqrt{2}W_t - \tau^\alpha\gamma^{(j-1)\alpha}\right\}.
$$

Consequently,

$$
\mathrm{P}(A) \leq 2\sum_{j=1}^{\infty}\mathrm{P}\left(\max_{\tau\gamma^{j-1}\leq t\leq\tau\gamma^j}\sqrt{2}W_t > \tau^\alpha\gamma^{(j-1)\alpha} - j\log[(1+q)\gamma] + \log(\epsilon\delta^2 q/\tau)\right)
$$

$$
\leq 2\sum_{j=1}^{\infty}\exp\left\{-\frac{1}{4\gamma^\alpha}(\tau^{\frac{\alpha}{2}}\gamma^{(j-1)\frac{\alpha}{2}} - b_j - \gamma^{\frac{\alpha}{2}})^2\right\}, \qquad (10.16)
$$

for $b_j = \tau^{-\frac{\alpha}{2}}\gamma^{-(j-1)\frac{\alpha}{2}}[j\log[(1+q)\gamma] - \log(\epsilon\delta^2 q/\tau)]$. Associating this inequality with (10.12) will lead to a lower bound on the difference between the Pickands constant and the truncated constant.

The last task is to find an upper bound for $\mathrm{E}(\mathcal{M}_0/\hat{\mathcal{S}}_0)$ in terms of $\mathrm{E}(\hat{\mathcal{M}}_0/\hat{\mathcal{S}}_0)$. Both terms are truncated to the interval $[-\tau, \tau]$. In the second term the maximization $\hat{\mathcal{M}}_0$ is conducted only over the approximating grid and not over the continuum as in the first term.

Clearly,

$$
\mathrm{E}(\mathcal{M}_0/\hat{\mathcal{S}}_0) \leq e^\epsilon\mathrm{E}(\hat{\mathcal{M}}_0/\hat{\mathcal{S}}_0) + \mathrm{E}(\mathcal{M}_0/\hat{\mathcal{S}}_0; \Delta(-\tau, \tau) > \epsilon).
$$

We would like to use (10.12) again to bound $E(\mathcal{M}_0/\hat{\mathcal{S}}_0; \Delta(-\tau, \tau) > \epsilon)$. For that we need a bound on the probability $P(\Delta(-\tau, \tau) > \epsilon)$.

Notice that $\kappa(-\tau, \tau) = \max\{\delta^\alpha, \tau^\alpha - (\tau - \delta)^\alpha\}$. Use the self-similarity in conjunction with Borell's inequality and the bound $E(\max_{0 \leq t \leq 1} \sqrt{2}W_t) \leq 1$ to deduce that

$$P(\Delta(-\tau, \tau) > \epsilon) \leq P(\Gamma(-\tau, \tau) > \epsilon - \kappa(-\tau, \tau))$$

$$\leq \frac{2\tau}{\delta} P\left(\sqrt{2}\delta^{\frac{\alpha}{2}} \sup_{0 \leq t \leq 1} W_t > \epsilon - \kappa(-\tau, \tau)\right)$$

$$\leq \frac{2\tau}{\delta} \exp\left\{-0.5\left(\frac{\epsilon - \kappa(-\tau, \tau)}{\sqrt{2}\delta^{\alpha/2}} - 1\right)^2\right\},$$

for all $\epsilon > \kappa(-\tau, \tau)$.

With the decrease in δ the exponential decay takes over. However, this bound is the more problematic one and the bottleneck in the application of the method. The error is determined by ϵ. The ratio between ϵ and $\delta^{\frac{\alpha}{2}}$ should still be large in order to ensure the probabilistic bound. This is attainable for larger values of α but becomes more and more difficult when α decreases. We were still able to maintain good accuracy for values of α in the range $1 \leq \alpha \leq 2$. Luckily, this is the range of more practical interest. However, extending the numerical evaluation of the constants to the lower half of the range of α values will require better methods for assuring accurate approximation of the process by a discrete approximation.

Another point to make is that the error analysis that we have just finished can be used in order to prove the uniform integrability of the right-hand side of (10.3). The point is that the integral in the denominator of the ratio between the maximum of the exponentiated process and its integral can be bounded from below by the Riemann sum times the exponent of $-\Lambda(t_0, t_1)$, where $[t_0, t_1]$ is the range of integration and maximization and $\Lambda(t_0, t_1) = \inf_{t_0 \leq t \leq t_1} Z_t - Z_t^\delta$. Incorporating this term in the proof will produce a simple generalization of the bound that was obtained for $E(\mathcal{M}(t_0, t_1)/\hat{\mathcal{S}}(t_0, t_1); A)$ and extend it to a bound on terms such as $E(\mathcal{M}(t_0, t_1)/\mathcal{S}(t_0, t_1); A)$. The rest of the argument is, more or less, unchanged.

Return to the evaluation of the Pickands constants. The bound on the error, that combines the lower and upper bound from truncation and the error from discretization, served as a basis for a simulation study for the computation of the Pickands constants in the range $1 < \alpha < 2$. Simulation of fractional Brownian motion is highly nontrivial, but there exists a vast body of literature on the topic. The fastest available algorithms simulate the process on an equispaced grid, by simulation of the (stationary) increment process of the fractional Brownian motion. For that the method of Davies and Harte [31] was used.

The truncation and discretization errors critically depend on α, but we choose τ and δ to be fixed throughout our experiments in order to be able to reuse the simulation design for all values of α in the range. The selection of τ and α was based on the worst case scenario in the given range, namely the case $\alpha = 1$ of the Brownian motion. Since τ and δ are fixed, our estimates for $\mathcal{H}_\alpha^\delta(\tau)$ are likely to be far off from $\mathcal{H}_\alpha$ for smaller values of α.

We estimate $\mathcal{H}_\alpha^\delta(\tau)$ using 1500 simulation replications, which took about 3 days to run for each value of α. The simulation was conducted for $\alpha = 14/20, 15/20, \ldots, 40/20$. However, as mentioned above, only values for $\alpha \geq 20/20$ can be trusted. A high-performance computing environment was used to run the experiments in parallel.

The parameters were selected so that the simulated error bounds from the previous section yield an error of approximately 3% for $\alpha = 1$. The most crucial parameter in the error analysis is ϵ. We note that a different ϵ can be used for the lower and upper bounds, and that ϵ may depend on α, so we take advantage of this extra flexibility to carefully select ϵ. For the upper bound we use $\epsilon = 0.005 + 0.025 \cdot (2 - \alpha)$, and for the lower bound we use $\epsilon = (0.005 + 0.025 \cdot (2 - \alpha))/3$. The values of the other parameters were $\tau = 128$ and $\delta = 1/2^{18}$.

The sampling standard deviation of our estimates for $\alpha = 1$ was 0.412, which corresponds to a sampling error of about 0.01, or 1% of the expected value. This relative error decreases, more or less linearly, until $\alpha = 1.85$ where it is equal to about 0.5% and then drops even sharper.

The outcomes of the simulation study are presented in Figure 10.1. The solid line gives the simulated values of $\mathcal{H}_\alpha^\delta(\tau)$ with a linear interpolation between points where the evaluation took place. The upper and lower upper bounds that result from the analysis of the discrepancy between $\mathcal{H}_\alpha$ and $\mathcal{H}_\alpha^\delta(\tau)$ are presented as broken lines. This error analysis is truncated at $\alpha = 1$. Simulated values below this point are less reliable.

A conjecture that was quoted before states that $\mathcal{H}_\alpha = 1/\Gamma(1/\alpha)$ [32]. The given conjecture is plotted as a dot-dashed line. It is clear that the conjecture does not fit the outcomes of the simulation experiment. This falls short of a mathematical proof. Yet, in a statistical analysis the null hypothesis that states that the conjecture is correct will be rejected with an extremely small p-value. Nonetheless, the conjectured formula can perhaps serve as a reasonable approximation for $\alpha \geq 1$, possibly with minor modifications to improve its fit to the computed line.

10.4   Enumerating the effect of local fluctuations

This chapter is the last in the book. It deals with a rather technical issue – not a very inspiring way to wrap up our story. By way of compensation we devote this section to presenting an open problem, not a solution. We initially discuss this problem in the context of constants produced by the fractional Brownian model and then extend the question to more general models.

In order to present the problem let us return to the example of the sequential probability ratio test, the example that opened the book. As part of the

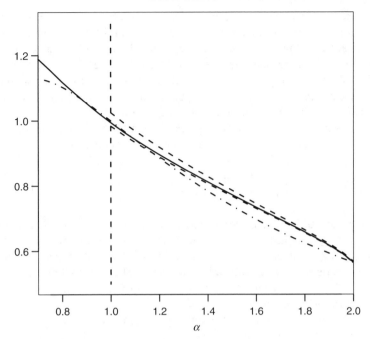

Figure 10.1   Estimate of the constants (solid line) and upper and lower bounds that are obtained by the error analysis (broken line). The function $1/\Gamma(1/\alpha)$ is also plotted (dot-dashed line). The upper and lower bounds are plotted only for $1 \leq \alpha \leq 2$.

analysis of this example we derived the overshoot modification for the approximation of the significance level. A rather lengthy analysis led to the representation of this modification in the form that employs infinite sums. The formula is given in (2.3). In particular, in the case of testing for a shift μ in the expectation of a standard normal measurement the term simplifies slightly. More important for practical applications is the fact that it has a good and simple approximation that is given in (2.4).

Relating the overshoot term presented in (2.4) to the topic of the current chapter we may identify that the given overshoot term is nothing but the constant $\mathcal{H}_1^\delta$ that is defined in (10.5). Specifically, $\delta = \mu^2/2$. This term, the Pickands constant for the discrete regular Brownian motion ($\alpha = 1$), reappeared in the approximating expression of the probability of extreme values in a random field.

Thinking of the constant as a function of δ we may identify some properties. The first property is the continuity, as $\delta \to 0$, of the function. This continuity is a corollary of the error analysis that we carried out in order to justify the approximation of $\mathcal{H}_\alpha$ by $\mathcal{H}_1^\delta(\tau)$ in general, which carries over of course to the case $\alpha = 1$. More is known regarding this limit. For example, one may obtain a one-sided Taylor expansion of the function [2]. In particular, since the value at 0 is known to be 1 and the derivative at zero is known, one may easily produce approximations of the overshoot function that are valid for small values of δ.

Consider the following question: can one extend this result to values of α other than $\alpha = 1$? Now formulate the question differently: we were able to compute the value of $\mathcal{H}_\alpha$ by the use of a truncated version of $\mathcal{H}_\alpha^\delta$ as an approximation. Can we reverse the process? Can we approximate $\mathcal{H}_\alpha^\delta$ on the basis of knowing the value of $\mathcal{H}_\alpha$ and, perhaps, other related quantities?

Here is a heuristic argument that proposes that the answer to the question is affirmative. Consider the difference between the two constants:

$$\mathcal{H}_\alpha(\tau) - \mathcal{H}_\alpha^\delta(\tau) = \mathrm{E}\left(\frac{\mathcal{M}_0}{\delta \hat{\mathcal{S}}_0}\right) - \mathrm{E}\left(\frac{\hat{\mathcal{M}}_0}{\delta \hat{\mathcal{S}}_0}\right) = \mathrm{E}\left(\left(e^{\log \mathcal{M}_0 - \log \hat{\mathcal{M}}_0} - 1\right) \times \frac{\hat{\mathcal{M}}_0}{\delta \hat{\mathcal{S}}_0}\right).$$

In the representation of the (truncated) Pickands constant we used a formula based on (10.4). The last expectation involves product of two random variables. One, $\hat{\mathcal{M}}_0/(\delta \hat{\mathcal{S}}_0)$ converges almost surely to $\mathcal{M}_0/\mathcal{S}_0$ that is defined over the continuous process and is uniformly integrable. The other random variable is bounded by $e^{\Delta(-\tau,\tau)} - 1$ and thus converges to zero almost surely when $\delta \to 0$, and is uniformly integrable for every finite τ.

We conjecture that the other random variable, multiplied by $\delta^{\frac{\alpha}{2}}$ converges in an appropriate matrix to a finite limit. Further, or so we hope, the converging sequence is uniformly integrable. If this is the case, and if it can be extended to the unrestricted process, then one may obtain the approximation:

$$\mathcal{H}_\alpha^\delta = \mathcal{H}_\alpha - \delta^{\frac{\alpha}{2}} \times \mathcal{D}_\alpha + o(\delta^{\frac{\alpha}{2}}),$$

with $\mathcal{D}_\alpha$ the anticipated limit. If everything falls in the right place and if there is a reliable method for the enumeration of $\mathcal{D}_\alpha$ then the outcome will be an extension of an approach that was extremely useful in producing approximations in cases where the limit of the local field is Gaussian and the increments are independent to cases where the limit still involves Gaussian processes but the increments are not independent.

And why stop in the one-dimensional limit? In the examples that we presented in the book the constants were explicit or, at least, had an expression that could be used for their evaluation. For example, for non-Gaussian limits of local processes we could in principle apply (2.3) and obtain the overshoot correction. This fact, which is caused in part by our selection bias in choosing the applications, may lead to the wrong impression that the constants are always easily evaluated.

This is not the case. For illustration return to the problem of detecting the emergence of a signal in a two-dimensional image, the problem that was discussed in Chapter 8. In that example, the image was scanned with a statistic that is based on a smooth kernel. The fact that we were able to obtain relatively simple looking formulae is a consequence of the smoothness of the kernel. If we had used a different kernel function, say the indicator of some two-dimensional set, then the situation would not have been that simple. A relatively simple formula may be obtained for a rectangle-shaped kernel. However, we do not know a formula for the case where the kernel is a circle.

I should make a reservation to the statement 'we do not know ...'. As a general phenomenon, one may obtain from the method that was advocated in this book a presentation of the relevant constant in the form of an expectation of a maximum of an exponentiated local field divided by the sum (or integral) of the exponentiated field. The local field in the case of a circle corresponds to local fluctuations of circles that diverge slightly from a given circle. Assuming asymptotic normality, one can even characterize the expectation and covariance functions of the limit local field. What we do not know is how to translate that presentation to a computable formula.

So here is the next question: can one find a general method for the evaluation of the expectation of the maximum over the sum in multivariate random fields? If not in general, can one do it for nonsmooth discrete Gaussian fields? Asking even less, can that be achieved for continuous Gaussian fields? And if so, can an approximation for the discrete case be developed on the basis of convergence to the smooth case?

As promised: many questions, no answers.

Appendix

Mathematical background

In this appendix we provide a short summary of relatively advanced material in probability that is assumed in the derivations that are given in the book. Most of the probabilistic material can be found in Durrett's book [33].

A.1   Transforms

Given a random variable X, the moment generating function that is associated with its distribution is

$$M_X(\theta) = E[e^{\theta X}] \,,$$

for all θ for which the function is finite. The function is non-negative and convex. The moments of the distribution coincide with the derivatives of the function evaluated at the origin. The moment generating function determines the distribution uniquely, provided that the origin is in the interior of the range of the function. The moment generating function of a sum of independent random variables is the product of their moment generating functions.

The cumulant generating function is the log of the moment generating function: $\psi(\theta) = \log M(\theta)$. We will typically call it the log-moment generating function. The cumulants are the derivative of the cumulant generating function, evaluated at the origin. The first cumulant is the expectation and the second cumulant is the variance. The log-moment generating function of the normal distribution with mean μ and variance σ^2 is $\theta\mu + \theta^2\sigma^2/2$.

A family of distribution measure is of an exponential class in the natural form if the density, with respect to an appropriate measure, can be written in the form $f_\theta(x) = h(x)\exp\{\theta \cdot t(x) - \psi(\theta)\}$. The sufficient statistic $t(x)$ and the

Extremes in Random Fields: A Theory and its Applications, First Edition. Benjamin Yakir.

natural parameter θ can be both vectors, in which case the product becomes an inner product. The expectation of $t(X)$ is given by the derivative $\psi'(\theta)$ and the variance is the second derivative $\psi''(\theta)$. A sum of i.i.d. random variables from an exponential family form an exponential family. Many, but not all, models of statistical interest are from an exponential family.

One may produce a family of distributions of an exponential class using the log-moment generating function $\psi(\theta)$. The log-likelihood ratio with respect to the baseline density, namely the log of the ratio between the density of the distribution associated with θ and the density associated with 0, is $\theta \cdot t(x) - \psi(\theta)$.

The large deviation rate for a sum of i.i.d. random variables may be obtained with the aid of the log-moment generating function:

$$\lim_{n \to \infty} n^{-1} \log P(S_n \geq xn) = -I(x) ,$$

where S_n is a sum of the random variables, $I(x) = \sup \{x\theta - \psi(\theta) : \theta \in \mathbb{R}\}$, and $x > E(X_1)$.

The characteristic function of a random variable X is the moment generating function evaluated at $\theta = it$. It is defined via:

$$\hat{\psi}(t) = Ee^{itX} = \int e^{itx} dF_X(x) .$$

Thus, it is the Fourier transform of the distribution of X. The definition extends to a random vector X by taking t to be a vector and replacing the product tX by the inner product $\langle t, X \rangle$. The characteristic function uniquely determines the distribution of X as is guaranteed by the inversion theorem:

$$\lim_{T \to \infty} \int_{-T}^{T} \frac{e^{-ita} - e^{-itb}}{it} \hat{\psi}(t) dt = P(a < X < B) + \frac{P(X = a) + P(X = b)}{2} .$$

When the distribution has a density with respect to the Lebesgue measure and the characteristic function is integrable then the inverse transform is

$$f(x) = \frac{1}{2\pi} \int_{-\infty}^{\infty} \hat{\psi}(t) e^{-itx} dt$$

The characteristic function of the normal distribution with mean μ and variance σ^2 is $\exp\{-t\mu - t^2\sigma^2/2\}$.

The error in the approximation of a characteristic function by a polynomial constructed with the moments of the random variable is bounded by:

$$\left| \psi(t) - \sum_{m=0}^{n} \frac{(it)^m E(X^m)}{m!} \right| \leq E \min \left(\frac{|tX|^{n+1}}{(n+1)!}, \frac{2|tX|^n}{n!} \right) .$$

In particular, if $E(X) = \mu$ and $Var(X) = \sigma^2 < \infty$ then $\psi(t) = 1 + it\mu - t^2\sigma^2/2 + o(t^2)$, and the error term is bounded by $6t^2\sigma^2$.

A.2 Approximations of sum of independent random elements

The specific local limit theorem is taken from [33]. The proof is essentially taken from Chapter 2 in Durrett's book that deals with local limit theorems. We give the proof here because in Section 5.4 we develop a slight generalization of the given theorem. Such a generalization is obtained by an appropriate modification to the original proof. Hence, we give the original proof as reference.

Theorem A.1 (Theorem 5.4 in [33]). *Let $X_1, X_2, \ldots$ be i.i.d. random variables. Assume that $\mathrm{E}(X_1) = 0$ and $\mathrm{Var}(X_1) = 1$. Let $\varphi(\theta) = \mathrm{E}(e^{i\theta X_1})$ be the characteristic function and assume that $|\varphi(\theta)| < 1$ for all $\theta \neq 0$. Then if $\delta > 0$ is fixed and if $x_n/\sqrt{n} \to x$, for a finite x, then*

$$\lim_{n \to \infty} \sqrt{n}\mathrm{P}(X_1 + \cdots X_n \in (x_n, x_n + \delta)) = \delta\phi(x) .$$

Proof. Let $\delta > 0$ and set $S_n = X_1 + \cdots X_n$. Consider the characteristic function of the Polya's distribution with density

$$h_0(y) = \frac{1}{\pi} \cdot \frac{1 - \cos(\delta y)}{\delta y^2} ,$$

which is of a bounded support and equal to:

$$\hat{h}_0(u) = \begin{cases} 1 - |u/\delta| & \text{if } |u| \leq \delta \\ 0 & \text{otherwise.} \end{cases}$$

Extend the density to a family of complex-valued functions by taking $h_\theta(y) = e^{i\theta y}h_0(y)$ and observe that $\hat{h}_\theta(u) = \hat{h}_0(u + \theta)$.
   We start by showing that for any given θ:

$$\lim_{n \to \infty} \sqrt{n}\mathrm{E}h_\theta(S_n - x_n) = \phi(x) \int h_\theta(y)dy , \qquad (A.1)$$

where ϕ is the density of the standard normal distribution. Indeed, from the inversion formula for characteristic functions in the presence of a density we get that:

$$h_0(x) = \frac{1}{2\pi} \int e^{-iux} \hat{h}_0(u)du$$

and therefore, by the change of variable $u = v + \theta$,

$$h_\theta(x) = e^{i\theta x}h_0(x) = \frac{1}{2\pi} \int e^{-i(u-\theta)x} \hat{h}_0(u)du = \frac{1}{2\pi} \int e^{-ivx} \hat{h}_\theta(v)dv .$$

Denote the distribution of $S_n - x_n$ by F_n and apply Fubini's theorem:

$$Eh_\theta(S_n - x_n) = \frac{1}{2\pi} \iint e^{-iuy} \hat{h}_\theta(u) du dF_n(y) = \frac{1}{2\pi} \iint e^{-iuy} dF_n(y) \hat{h}_\theta(u) du .$$

The innermost integral corresponds to the characteristic function of $S_n - x_n$, evaluated at $-u$, hence

$$= \frac{1}{2\pi} \int [\varphi(-u)]^n e^{iux_n} \hat{h}_\theta(u) du .$$

In order to show that the limit of the given integral is equal to the right-hand side of (A.1) we consider three regions. The first region is the region $[-\epsilon, \epsilon]$, for an appropriate $\epsilon > 0$, with the property that $|\varphi(-u)| \leq \exp(-u^2/4)$ over the region. The second region is $[-M, M] \setminus [-\epsilon, \epsilon]$, where $[-M, M]$ contains the support of h_θ. The last region is $\mathbb{R} \setminus [-M, M]$, over which the integrand is equal to 0.

The last region does not contribute to the integral. The contribution of the second region is $o(n^{-\frac{1}{2}})$ small since it is bounded by $(M/\pi)\eta^n$, for $\eta = \sup_{\epsilon \leq |u| \leq M} |\varphi(-u)| < 1$. For the first region we have, after multiplying by $\sqrt{n}$ and changing the variable to $v = u\sqrt{n}$:

$$\frac{\sqrt{n}}{2\pi} \int_{-\epsilon}^{\epsilon} [\varphi(-u)]^n e^{iux_n} \hat{h}_\theta(u) du = \frac{1}{2\pi} \int_{-\epsilon\sqrt{n}}^{\epsilon\sqrt{n}} [\varphi(-v/\sqrt{n})]^n e^{ivx_n/\sqrt{n}} \hat{h}_\theta(v/\sqrt{n}) dv .$$

The integrand converges, for each fixed v, to $\exp(-v^2/2 + ivx)\hat{h}_\theta(0)$. Application of the dominated convergence theorem will give:

$$\xrightarrow[n\to\infty]{} \frac{1}{2\pi} \int_{-\infty}^{\infty} \exp(-v^2/2 + ivx)\hat{h}_\theta(0) dv = \phi(x)\hat{h}_\theta(0) = \phi(x) \int h_\theta(y) dy .$$

The last equality follows from the definition of the Fourier transform. This completes the proof of (A.1), which we will use next in order to prove the statement of the theorem.

To that end, we consider two sequences of measures on the real line. The first is the measure

$$\mu_n(A) = \sqrt{n} P(S_n - x_n \in A) ,$$

that converges to the measure $\phi(x)\mu(A)$, for μ the Lebesgue measure. The other measure is the probability measure

$$\nu_n(A) = \frac{1}{\alpha_n} \int_A h_0(y)\mu_n(dy) ,$$

for $\alpha_n = \sqrt{n} E h_0(S_n - x_n \in A)$. From (A.1) it follows, for $\theta = 0$, that $\alpha_n \to \phi(x)$ and more generally that:

$$\int e^{i\theta y} d\nu_n(y) = \frac{1}{\alpha_n} \sqrt{n} E h_\theta(S_n - x_n) \xrightarrow[n\to\infty]{} \int e^{i\theta y} h_0(y) dy .$$

Consequently, ν_n converges in distribution to the Polya's distribution. For the final move we apply the likelihood ratio identity:

$$\frac{1}{\alpha_n}\mu_n([0,\delta]) = \int \frac{1_{[0,\delta]}(y)}{h_0(y)} d\nu_n(y) \to_{n\to\infty} \int \frac{1_{[0,\delta]}(y)}{h_0(y)} h_0(y) dy = \delta ,$$

which completes the proof of the theorem.   □

The classical method for dealing with the distribution of the sum of independent random variables uses approximations of the characteristic function of the sum. The central limit theorem in its general form was originally proved this way as well as the Berry–Esseen theorem. This theorem states that for a standardized sum of i.i.d. random variables Z_n:

$$\sup_{x\in\mathbb{R}} |P(Z_n \le x) - \Phi(x)| \le C\frac{\mu_3}{\sigma^3\sqrt{n}} ,$$

for μ_3 the centered third moment of an increment in the sum, σ^2 the variance, and C some universal constant.

With more conditions the result may be refined. For example, if we assume the existence of a bounded and continuous density and some moment conditions then one may be able to prove the following.

Theorem A.2 (Theorem 19.2 in [11]). *Let X_n be an i.i.d. sequence of random vectors in $\mathbb{R}^d$, with common zero mean and positive-definite variance matrix V. Assume that $\rho_s = \mathrm{E}\|X_1\|^s < \infty$ for some integer $s > 3$ and that the characteristic function of X_1 belongs to L_p for some $p \ge 1$. Then a bounded and continuous density q_n for the re-normalized sum $(X_1 + \cdots X_n)/\sqrt{n}$ exists for $n \ge p$ and this density has an expansion in terms of the normal density and vanishingly small polynomial corrections. The error in the expansion is of order $o(n^{-(s-2)/2})$, uniformly over compact regions. Some control is given on the tail behavior of the approximation.*

Here is yet another generalization that does not require densities. A distribution Q on $\mathbb{R}^d$ is said to satisfy Cramér's condition if:

$$\limsup_{\|t\|\to\infty} |\psi(t)| < 1 ,$$

where $\psi(t)$ is the characteristic function of Q. Let $\sum_{r=0}^{k-2} n^{-\frac{r}{2}}P(-\Phi_\Sigma, \{\chi_\nu\})$ a higher order approximation of a distribution, with respect to the Gaussian distribution, be given in terms of the cumulants of the distribution. Define

$$M_s(f) = \begin{cases} \sup_{x\in\mathbb{R}^d}(1 + \|x\|^s)^{-1}|f(x)| & s > 0 , \\ \sup_{x,y\in\mathbb{R}^d}|f(x) - f(y)| & s = 0 . \end{cases}$$

and let

$$\overline{\omega}_f(2e^{-cn}, \Phi_\Sigma) = \int \sup_{\{y:|y-x|\le 2e^{-cn}\}} |f(y) - f(x)|\Phi_\Sigma(dx) .$$

Theorem A.3 (Theorem 20.1 in [11]).   *Let X_n be an i.i.d. sequence of random vectors in $\mathbb{R}^d$, whose common distribution Q_1 satisfies Cramér's condition. Assume that Q_1 has mean zero and a finite kth absolute moment for some integer $k \geq 3$. Let Σ denote the covariance matrix of Q_1 and χ_v its vth cumulant $(3 \leq |v| \leq k)$. Then for every real-valued, Borell-measurable function $f : \mathbb{R}^d \rightarrow \mathbb{R}$ satisfying $M_s(f) < \infty$, for some $0 \leq s \leq k$, one has that:*

$$\left| \int fd \left(Q_n - \sum_{r=0}^{k-2} n^{-\frac{r}{2}} P(-\Phi_\Sigma, \{\chi_v\}) \right) \right| \leq M_s(f)\delta(n) + c(k,d)\overline{\omega}_f(2e^{-cn}, \Phi_\Sigma),$$

where Q_n is the distribution of $n^{-\frac{1}{2}} \sum_{i=1}^n X_i$, Φ_Σ is the multivariate normal distribution with zero mean and variance-covariance matrix Σ, c is an absolute constant, $c(k,d)$ constants that depend on the dimension of the space and the level of approximation, and $\delta(n) = o(n^{-(k-2)/2})$. All terms that are not otherwise specified do not depend on f.

A.3   Concentration inequalities

Concentration inequalities provide bounds, typically of exponential or super-exponential rate, on the probability of obtaining values away from the expectation. A useful such bound is the Chernoff bound, which is obtained from the analysis of the log-moment generating function of the random variable. A formulation of a Chernoff-type of bound on extreme positive values, applied to a sum of bounded random variables, states the following.

Theorem A.4   *Let X_i be independent random variables satisfying $X_i \leq E(X_i) + M$, for $1 \leq i \leq n$, let $X = \sum_{i=1}^n X_i$, and denote $\sigma^2 = \mathrm{Var}(X)$. Then*

$$P(X - E(X) \geq z\sigma) \leq e^{-\frac{1}{2}z^2 \frac{1}{1+zM/(3\sigma)}}.$$

A similar result holds for extreme negative values.

A concentration inequality for marginal distributions, in combination with a chaining argument, can be used in order to produce a concentration inequality for a random variable that is produced by the maximization of a random field. An example is Fernique's inequality that is stated as Theorem 3.1 in the context of a Gaussian random field. A generalization of the argument can be used in order to produce Fernique's type of inequalities for random fields that are not Gaussian. Such theory is developed, for example, in [17].

Theorem A.5 ([17], Theorem 3.1).   *For each fixed positive integer n, there exist universal constants C_n and C'_n, depending only on n, such that if X defined on I is a separable sub-nth-Gaussian chaos field with respect to the pseudo-metric δ, then for each $t_0 \in I$ and $s \geq 0$,*

$$P\left(\sup_{t \in I} |X_t - X_{t_0}| > C_n M + s C' D \right) \leq 2e^{-\frac{1}{2}s^{2/n}}$$

where $M = \int_0^\infty [\log N_\delta(\epsilon)]^{n/2} d\epsilon$ and D is the diameter of I under δ. In addition, if D and M are finite, then X is almost-surely bounded.

A.4   Random walks

A random walk is a process produced by partial sums of i.i.d. random variables S_n. Statements about the distribution associated with the process are of interest. For example, the law of large numbers states that if the expectation of an increment is finite then $(1/n)S_n \to E(X_1)$, almost surely.

Results may be obtained in relation to a random walk stopped by a stopping time. For example, Wald's identity states that if the expectation of an increment is finite and if the expectation of the stopping rule N is finite then $E(S_N) = E(X_1)E(N)$.

A useful inequality that can be proved using stopping times is Kolmogorov's maximal inequality that states that if the increments are centered and the variance of an increment σ_i^2 is finite then

$$P\left(\max_{1 \leq k \leq n} |S_k| \geq x\right) \leq \frac{1}{x^2} \sum_{i=1}^n \sigma_i^2 . \tag{A.2}$$

In the book, when we discuss the asymptotic distribution of an overshoot for a random walk stopped by a boundary we refer to a theorem that appears in Feller's book [32].

Theorem A.6 (Theorem 4 in [34]).   *Consider the process S_n of partial sums of independent random variables. Let $\tau_n = P(S_1 \leq 0, \ldots, S_{n-1} \leq 0, S_n > 0)$ and let $p_n = P(S_1 > 0, \ldots, S_{n-1} > 0, S_n > 0)$. Define the generating functions $\tau(s) = \sum_{n=1}^\infty \tau_n s^n$ and $p(s) = \sum_{n=1}^\infty p_n s^n$. Then*

$$\log p(s) = -\log(1 - \tau(s)) = \sum_{n=1}^\infty \frac{s^n}{n} P(S_n > 0) .$$

A.5   Renewal theory

Renewal theory deals with the long-run characteristics of a system with cycles of random length. These lengths are i.i.d.

Let S_n be the time at which the nth cycle ended, with $S_0 = 0$. The renewal measure is given by

$$U(A) = \sum_{n=0}^\infty P(S_n \in A)$$

and it corresponds to the expected number of cycle renewals that occurred in the event A. Blackwell's renewal theorem states that if the distribution of the cycle length is nonarithmetic then:

$$U([t, t + h)) \longrightarrow_{t \to \infty} h/\mu ,$$

where μ is the expected length of a cycle. A renewal process that initiates with a cycle that has the distribution

$$G(t) = \frac{1}{\mu} \int_0^t [1 - F(y)]dy ,$$

is stationary. Here F is the distribution of a typical cycle and G corresponds to the distribution of the remaining time in a cycle that covers time t.

In the book we use the renewal theory to justify the evaluation of the distribution of the overshoot in a stopped random walk. This theory can be used also in order to produce the expectation of this overshoot. It follows that the asymptotic expectation of the overshoot is equal to $E(Y_1^2)/[2E(Y_1)]$, where Y_1 is the first ladder height. Based on Wald's identity and on this approximation we get for the one-sided sequential probability-ratio test that:

$$E_g(\ell_1)E_g(N_x) = E_g(\ell_{N_x}) \approx x + E(Y_1^2)/[2E(Y_1)] .$$

An approximation for the expectation of the stopping time follows by dividing through by the expectation of an increment of the log-likelihood ratio process.

Nonlinear renewal theory deals with the theory associated with processes that can be represented as the sum of a random walk and a slowly changing perturbation. This theory covers cases such as random walks that are stopped by a curved, rather than linear, boundary as well as the case where a mixture-type likelihood ratio is used. The conclusion of the theory is essentially that the asymptotic distribution of the overshoot from the boundary is not affected by the perturbation. On the other hand, the expected time to stopping is modified by the perturbation:

$$E_g(N_x) \approx (x + E(Y_1^2)/[2E(Y_1)] - E_g\eta)/E_g(\ell_1) ,$$

where $E_g\eta$ is the expectation of the limit in distribution of the perturbation.

A.6   The Gaussian distribution

The standard normal distribution has the density $\phi(x) = (2\pi)^{-\frac{1}{2}} \exp\{-x^2/2\}$ over the real line and the cumulative distribution function $\Phi(x) = \int_{-\infty}^x \phi(z)dz$. The cumulative distribution function does not have a closed form, but a good asymptotic approximation, for large values of x, can be obtained with the aid of Mill's ratio:

$$\frac{x}{x^2 + 1}\phi(x) \leq \Phi(x) \leq \frac{1}{x}\phi(x) .$$

Modifications can be used to obtain better evaluations near the origin.

A random variable X has normal distribution if it is a linear transformation of a standard normal random variable: $X = aZ + b$, for Z standard normal.

In general, a random element is Gaussian if any linear functional applied to it is normally distributed.

A Gaussian vector is characterized by the vector of expectations of the components and the matrix of covariances between components. Decompose a Gaussian random vector $X = (X_1, X_2)'$ into sub-vectors. Also, decompose accordingly the vector of expectations and matrix of covariances:

$$\mu = \begin{pmatrix} \mu_1 \\ \mu_2 \end{pmatrix}, \quad \Sigma = \begin{pmatrix} \Sigma_{11} & \Sigma_{12} \\ \Sigma_{21} & \Sigma_{22} \end{pmatrix}.$$

The conditional distribution of X_1, given the value of X_2 – denoted the regression of X_1 on X_2 – is Gaussian. If Σ_{22} is of full rank then

$$\mathrm{E}(X_1|X_2) = \mu_1 + \Sigma_{12}\Sigma_{22}^{-1}(X_2 - \mu_2), \quad \mathrm{Var}(X_1|X_2) = \Sigma_{11} - \Sigma_{12}\Sigma_{22}^{-1}\Sigma_{21}.$$

In particular, the conditional covariance structure of X_1 does not depend on the values of X_2.

Two central results that are used in the analysis of extremes in a centered Gaussian random field are Borell's inequality and Slepian's inequality. The former gives a bound on the extreme tail and the latter makes a comparison between two random fields with one being more inter-correlated than the other.

Borell's inequality is similar to Fernique's inequality (Theorem 3.1).

Theorem A.7 (Borell's inequality). *Let $\{X_t : t \in T\}$ be a centered Gaussian field with almost surely bounded realizations. Let $\|X\| = \sup_{t \in T} X_t$ and let $\sigma^2 = \sup_{t \in T} \mathrm{Var}(X_t) < \infty$. Then $\mathrm{E}\|X\| < \infty$ and for all $x > 0$:*

$$\mathrm{P}(|\|X\| - \mathrm{E}\|X\|| > x) \leq 2e^{-\frac{1}{2}x^2/\sigma^2}$$

Bounds on the term $\mathrm{E}\|X\|$ may be obtained via considerations of entropy.

Slepian's inequality introduces another centered random field $\{Y_t : t \in T\}$ with the same variance: $\mathrm{Var}(Y_t) = \mathrm{Var}(X_t)$, for all $t \in T$.

Theorem A.8 (Slepian's inequality). *If $\mathrm{Cov}(X_t, X_s) \geq \mathrm{Cov}(Y_t, Y_s)$, for all $(s, t) \in T \times T$, then for all x:*

$$\mathrm{P}(\|X\| > x) \leq \mathrm{P}(\|Y\| > x).$$

Laplace approximation of an integral.

A.7   Large sample inference

Let $f_\theta(x)$ be the density of a random element X in a parametric family parameterized by θ, for θ in a finite dimension and 'nice' set. The likelihood is the density evaluated at the observed value of the element and considered as a function of the parameter. Assume that the density is smooth as a function of θ. The maximum likelihood estimator $\hat{\theta}$ is obtained by the maximization of the likelihood function or, equivalently, by the maximization of the log-likelihood: $\log f_\theta(X)$. Under appropriate regularity conditions this maximization can be obtained by the solution of the normal equation:

$$\frac{\partial}{\partial\theta}\log f_\theta(X) = \dot{f}_\theta(X)/f_\theta(X) = 0 .$$

The derivative of the log-likelihood itself, evaluated at the actual parameter θ, is called the score function. This function plays a crucial role in the large sample theory for the maximum likelihood estimator in the regular parametric setting. In particular, the expectation of the score function is equal to 0 and the variance of the score, which is also called the Fisher information index, is equal to the negative of the expectation of the derivative of the score:

$$I(\theta) = \mathrm{Var}_\theta\left(\frac{\partial}{\partial\theta}\log f_\theta(X)\right) = -\mathrm{E}_\theta\left(\frac{\partial^2}{\partial^2\theta}\log f_\theta(X)\right) .$$

In the special case of i.i.d. observations we obtain that the Fisher information for n observations is n times the Fisher information for a single observation:

$$I_n(\theta) = nI_1(\theta) .$$

In this setting the score function is asymptotically normal. Moreover, by the application of a Taylor expansion of the score function about its maximizing value, and by the δ-method we get that the asymptotic distribution of the maximum likelihood estimator is normal. The asymptotic expectation is θ and the asymptotic variance is the inverse of the Fisher information index.

The score function is a useful method for constructing statistical hypothesis tests. The score statistic is obtained by the evaluation of the function at the value of the parameter that is being tested. If the null hypothesis is composite then the evaluation can be taken as an estimate of the parameter, where the estimator is constraint to the null hypothesis. It turns out that in the large sample setting the resulting test is asymptotically as efficient as the standard generalized likelihood-ratio test.

A.8   Integration

Some facts regarding integration and convergence are given in the following.

Theorem A.9 (Fubini's theorem). *Suppose $X \times Y$ is a complete measure space equipped with the product measure $d(x, y) = dx \times dy$ and suppose that $f(x, y)$ is a measurable and integrable function then:*

$$\iint_{X\times Y} f(x, y)d(x, y) = \int_X\left(\int_Y f(x, y)dy\right)dx = \int_Y\left(\int_X f(x, y)dx\right)dy .$$

Let $\{f_n\}$ be a sequence of real valued measurable functions over a measure space. Assume that $f_n(x)\to_{n\to\infty}f(x)$, for almost all x.

Theorem A.10 (Dominated convergence theorem). *If $|f_n(x)| \le g(x)$ and g is a measurable function with $\int g(x)dx < \infty$ then $\int |f(x)|dx < \infty$. Moreover, we have that $\lim_{n\to\infty}\int |f_n(x) - f(x)|dx = 0$. In particular, $\lim_{n\to\infty}\int f_n(x)dx = \int f(x)dx$.*

A translation of this theorem to the language of random variables will imply from the point-wise convergence $X_n(\omega) \to X(\omega)$ and the relation $|X_n| \le Y$, with $EY < \infty$, the convergence of the expectations: $\lim_{n\to\infty}EX_n = EX$. For the convergence of the expectations to hold it is sufficient that $X_n \to X$ in probability.

A more general concept is the concept of uniform integrability: we say that a collection of random variables $\{X_n\}$ is uniformly integrable if for all $\epsilon > 0$ one can find a universal x such that:

$$E(|X_n|; |X_n| > x) \le \epsilon, \quad \text{for all } X_n \text{ in the collection.}$$

This concept is tightly linked to the concept of convergence of expectations. Given a sequence of random variables $\{X_n\}$ that converges in probability to a random variable X then $\lim_{n\to\infty}E|X_n - X| = 0$ if, and only if, the collection $\{X_n\}$ is uniformly integrable.

A.9   Poisson approximation

The Poisson distribution serves as an approximation to the number of occurrences of independent rare events. Consider, for example, a binomial random variable X_n associated with the count of the number of successes among n trials with probability of success p_n in each. It is easy to show that if $np_n \to \lambda$ then $\lim_{n\to\infty}P(X_n = x) = e^{-\lambda}\lambda^x/x!$, for any non-negative integer x.

More sophisticated theorems will establish a similar type of convergence when the events do not have identical probabilities and/or are weakly dependent. For example, in the book we use the following theorem.

Theorem A.11 (Theorem 1 in [9]). *Let $\{X_i : i \in I\}$ be Bernoulli random variables. Set $\hat{W} = \sum_i X_i$ and $\lambda = \sum_i P(X_i = 1) \in (0, \infty)$. Associate with each i a neighborhood of dependence $I_i \subset I$. Define:*

$$b_1 = \sum_{i\in I}\ \sum_{j\in I_i \setminus \{i\}} P(X_i = 1)P(X_j = 1)$$

$$b_2 = \sum_{i\in I}\ \sum_{j\in I_i \setminus \{i\}} P(X_i = 1, X_j = 1)$$

$$b_3 = \sum_{i\in I} E\left[|E(X_i|\sigma\{X_j : j \notin I_i\}) - E(X_i)|\right]$$

Then

$$2 \sup_A |P(\hat{W} \in A) - P(W \in A)| \le 2(b_1 + b_2 + b_3) \,,$$

where W is a Poisson random variable with expectation λ. *Also,*

$$|P(\hat{W} = 0) - e^{-\lambda}| \le (1 \wedge \lambda^{-1})(b_1 + b_2 + b_3) \,.$$

A.10   Convexity

Theorem A.12 (Jensen's inequality).   *Let X be a random variable with finite expectation and φ be a convex function. Then:* $\varphi(\mathrm{E}(X)) \le \mathrm{E}(\varphi(X))$. *If the function is strictly convex the equality holds only if the random variable is constant in probability 1.*

It follows from Jensen's inequality, since $-\log(x)$ is convex, that

$$\mathrm{E}[\log\{g(X)/f(X)\}] = \int \log\{g(x)/f(x)\}f(x)dx$$

$$< \log \int \{g(x)/f(x)\}f(x)dx = 0.$$

By symmetry, $I = \mathrm{E}_g[\log\{g(X)/f(X)\}] = -\mathrm{E}_g[\log\{f(X)/g(X)\}] > 0.$

References

[1] Seigmund D.O. *Sequential Analysis: Tests and Confidence Intervals*. Springer-Verlag, New York (1985).

[2] Chang J., Peres Y. Ladder heights, Gaussian random walks and the Riemann zeta function. *Ann. Probab.* **25**, 787–802 (1997).

[3] Seigmund D.O., Yakir B. *The Statistics of Gene Mapping*. Springer-Verlag, New York (2010).

[4] Adler R.J., Taylor J.E. *Random Fields and Geometry*. Springer, New York (2007).

[5] Piterbarg V.I. *Asymptotic Methods in the Theory of Gaussian Processes and Fields*. American Mathematical Society, Providence, RI (1996).

[6] Kuriki S., Takemura A. Volume of tubes and distribution of the maxima of Gaussian random fields. In *Selected Papers on Probability and Statistics*, American Mathematical Society Translations Series 2, Volume 227. American Mathematical Society, Providence, RI, 25–48 (2009).

[7] Aldous D. *Probability Approximations via the Poisson Clumping Heuristic*. Springer-Verlag, New York (1989).

[8] Lin Z., Bai Z. *Probability Inequalities*. Science Press, Beijing and Springer-Verlag, Berlin (2010).

[9] Arratia R., Goldstein L., Gordon L. Two moments suffice for poisson approximations: the Chen-Stein method. *Ann. Probab.* **17**, 9–25 (1989).

[10] Senatov V.V. *Normal Approximation: New Results, Methods and Problems*. VSP VB, Utrecht (1998).

[11] Bhattacharya R.N., Rao R.R. *Normal Approximation and Asymptotic Expansions*. John Wiley & Sons, Ltd, New York (1976).

[12] Nardi Y., Siegmund D.O., Yakir B. The distribution of maxima of approximately Gaussian random fields. *Ann. Statist.* **36**, 1375–1403 (2008).

[13] Massart P. The tight constant in the Dvoretzky-Kiefer-Wolfowitz Inequality. *Ann. Probab.* **18**, 1269–1283 (1990).

[14] Hu I. A uniform bound for the tail probability of Kolmogorov-Smirnov statistics. *Ann. Statist.* **13**, 821–826 (1985).

[15] Wei F., Dudley R.M. Two-sample Dvoretzky-Kiefer-Wolfowitz inequalities. *Statist. Probab. Lett.* **82**, 636–644 (2012).

Extremes in Random Fields: A Theory and its Applications, First Edition. Benjamin Yakir.
© 2013 by Higher Education Press. All rights reserved. Published 2013 by John Wiley & Sons, Ltd.

[16] Peacock J.A. Two-dimensional goodness-of-fit testing in astronomy. *Monthly Notices of the Royal Astronomical Society* **202**, 615–627 (1983).

[17] Viens F.G., Vizcarra A.B. Supremum concentration inequality and modulus of continuity for sub- nth chaos processes. *J. Funct. Anal.* **248**, 1–26 (2007).

[18] Seigmund D.O., Yakir B. Tail probabilities for the null distribution of scanning statistics. *Bernoulli* **6**, 191–213 (2000).

[19] Zhang N.R., Siegmund D.O., Ji. H., Li J. Detecting simultaneous change-points in multiple sequences. *Biometrika* **97**, 631–645 (2010).

[20] Seigmund D.O., Yakir B., Zhang N.R. Tail approximations for maxima of random fields by likelihood ratio transformations. *Sequential Analysis* **29**, 245–262 (2010).

[21] Seigmund D.O., Yakir B., Zhang N.R. Detecting simultaneous variant intervals in aligned sequences. *Ann. Appl. Statist.* **5**, 645–668 (2011).

[22] Seigmund D.O., Yakir B., Zhang N.R. The false discovery rate for scanning statistics. *Biometrika* **98**, 979–985 (2011).

[23] Shiryayev A.N. On optimum methods in quickest detection problems. *Theory Probab. Appl.* **8**, 22–46 (1963).

[24] Pollak M. Optimal detection of a change in distribution. *Ann. Statist.* **13**, 206–227 (1985).

[25] Siegmund D.O., Yakir B. Minimax optimality of the Shiryayev-Roberts change-point detection rule. *J. Stat. Plan. Infer.* **138**, 2815–2825 (2008).

[26] Siegmund D.O., Yakir B. Detecting the emergence of a signal in a noisy image. *Statistics and Its Interface* **1**, 3–12 (2008).

[27] Taqqu M.S., Willinger W., Sherman R. Proof of a fundamental result in self-similar traffic modeling. *Comput. Commun. Rev.* **27**, 5–23 (1997).

[28] Piterbarg V.I. Large deviations of a storage process with fractional Brownian motion as input. *Extremes* **4**, 147–164 (2001).

[29] Dieker A.B., Yakir B. On asymptotic constants in the theory of Gaussian processes. arXiv:1206.5840 (2012).

[30] Adler R.J. An Introduction to Continuity, Extrema, and Related Topics for General Gaussian Processes Institute of Mathematical Statistics, Hayward, CA (1990).

[31] Davies R.B., Harte D.S. Tests for Hurst effect. *Biometrika* **74**, 95–102 (1987).

[32] Dęicki K., Mandjes M. Open problems in Gaussian fluid queueing theory. *Queueing Syst.* **68**, 267–273 (2011).

[33] Durrett R. *Probability: Theory and Examples* (2nd Edition). Duxbury Press, Belmont, CA (1995).

[34] Feller W. *An Introduction to Probability Theory and its Applications* (2nd Edition). John Wiley & Sons, Ltd, New York (1971).

Index

Average run length 78, 85, 157,
    161–2

Bayesian 145, 148–9, 157, 161–2
Bernoulli 114, 116, 121, 219
Berry-Esseen theorem 11, 60–61,
    68, 139, 213
Binomial 74–5, 96, 110–112, 114,
    121, 219
Boundary 29, 57, 68, 81, 91, 93,
    96, 100
Brownian
  bridge 107
  fractional 189, 191–2, 194–6,
        199, 201, 203–4
  motion 24, 35–8, 67, 167,
        188–9, 195, 204–5

Central Limit Theorem 6, 8,
        10–11, 44, 87, 95, 150,
        213
  local 10, 22, 59, 60, 87–9, 91–2,
        95, 149, 178, 211
Chaining argument 111–113,
        175–6, 200, 214
Change-point 79–81, 83, 144–5,
        147–8, 161–3
  sequential detection 13, 71–2,
        85, 143–6, 157,
        161–3, 166
Characteristic function 96–8, 100,
        210–213
Chi-square distribution 39, 129,
        150

Clumping heuristic, Poisson 40,
    187–8
Covariance 9–10, 30–31, 34–5,
    48–9, 53, 63–6,
    116–117, 135–7, 159,
    171, 180–182, 191–4,
    214
  structure 9, 107, 115, 122,
        167–8, 188–9,
        217

Dependence
  long-range 167, 169, 186–8
  neighborhood 75–6, 82, 219
  short-range 174–5
Discretization 8, 43, 46–7, 53–4,
    111, 121, 132, 175,
    194–6, 198–200
Distributions
  alternative 7–8, 11, 18–20,
        24–5, 38, 53, 78–9,
        105, 110, 114–119,
        122, 131–4, 146–9,
        159–63, 167, 177
  baseline 4, 145–6, 210
  empirical 105–9, 119–20, 123
  heavy-tail 168, 186, 188
  marginal 8, 29–30, 34, 43–6,
        110–111, 119,
        131–2, 148–9,
        173–4
  null 7, 16–8, 24–5, 27, 38, 53,
        78, 105, 115, 140,
        147–8, 156, 159–60

Extremes in Random Fields: A Theory and its Applications, First Edition. Benjamin Yakir.
© 2013 by Higher Education Press. All rights reserved. Published 2013 by John Wiley & Sons, Ltd.

WILEY SERIES IN PROBABILITY AND STATISTICS
Established by Walter A. Shewhart and Samuel S. Wilks

Editors: *David J. Balding, Noel A. C. Cressie, Garrett M. Fitzmaurice, Harvey Goldstein, Iain M. Johnstone, Geert Molenberghs, David W. Scott, Adrian F. M. Smith, Ruey S. Tsay, Sanford Weisberg*

Editors Emeriti: *Vic Barnett, J. Stuart Hunter, Joseph B. Kadane, Jozef L. Teugels*

The *Wiley Series in Probability and Statistics* is well established and authoritative. It covers many topics of current research interest in both pure and applied statistics and probability theory. Written by leading statisticians and institutions, the titles span both state-of-the-art developments in the field and classical methods.

Reflecting the wide range of current research in statistics, the series encompasses applied, methodological and theoretical statistics, ranging from applications and new techniques made possible by advances in computerized practice to rigorous treatment of theoretical approaches.

This series provides essential and invaluable reading for all statisticians, whether in academia, industry, government, or research.

† ABRAHAM and LEDOLTER · Statistical Methods for Forecasting
 AGRESTI · Analysis of Ordinal Categorical Data, *Second Edition*
 AGRESTI · An Introduction to Categorical Data Analysis, *Second Edition*
 AGRESTI · Categorical Data Analysis, *Second Edition*
 ALTMAN, GILL, and McDONALD · Numerical Issues in Statistical Computing for the
  Social Scientist
 AMARATUNGA and CABRERA · Exploration and Analysis of DNA Microarray and
  Protein Array Data
 ANDEĚL · Mathematics of Chance
 ANDERSON · An Introduction to Multivariate Statistical Analysis, *Third Edition*
* ANDERSON · The Statistical Analysis of Time Series
 ANDERSON, AUQUIER, HAUCK, OAKES, VANDAELE, and WEISBERG·
  Statistical Methods for Comparative Studies
 ANDERSON and LOYNES · The Teaching of Practical Statistics
 ARMITAGE and DAVID (editors) · Advances in Biometry
 ARNOLD, BALAKRISHNAN, and NAGARAJA · Records
* ARTHANARI and DODGE · Mathematical Programming in Statistics
* BAILEY · The Elements of Stochastic Processes with Applications to the Natural
  Sciences
 BAJORSKI · Statistics for Imaging, Optics, and Photonics
 BALAKRISHNAN and KOUTRAS · Runs and Scans with Applications
 BALAKRISHNAN and NG · Precedence-Type Tests and Applications
 BARNETT · Comparative Statistical Inference, *Third Edition*
 BARNETT · Environmental Statistics
 BARNETT and LEWIS · Outliers in Statistical Data, *Third Edition*
 BARTHOLOMEW, KNOTT, and MOUSTAKI · Latent Variable Models and Factor
  Analysis: A Unified Approach, *Third Edition*

* Now available in a lower priced paperback edition in the Wiley Classics Library.
† Now available in a lower priced paperback edition in the Wiley–Interscience Paperback Series.

BARTOSZYNSKI and NIEWIADOMSKA-BUGAJ · Probability and Statistical Inference, *Second Edition*

BASILEVSKY · Statistical Factor Analysis and Related Methods: Theory and Applications

BATES and WATTS · Nonlinear Regression Analysis and Its Applications

BECHHOFER, SANTNER, and GOLDSMAN · Design and Analysis of Experiments for Statistical Selection, Screening, and Multiple Comparisons

BEIRLANT, GOEGEBEUR, SEGERS, TEUGELS, and DE WAAL · Statistics of Extremes: Theory and Applications

BELSLEY · Conditioning Diagnostics: Collinearity and Weak Data in Regression

† BELSLEY, KUH, and WELSCH · Regression Diagnostics: Identifying Influential Data and Sources of Collinearity

BENDAT and PIERSOL · Random Data: Analysis and Measurement Procedures, *Fourth Edition*

BERNARDO and SMITH · Bayesian Theory

BHAT and MILLER · Elements of Applied Stochastic Processes, *Third Edition*

BHATTACHARYA and WAYMIRE · Stochastic Processes with Applications

BIEMER, GROVES, LYBERG, MATHIOWETZ, and SUDMAN · Measurement Errors in Surveys

BILLINGSLEY · Convergence of Probability Measures, *Second Edition*

BILLINGSLEY · Probability and Measure, *Anniversary Edition*

BIRKES and DODGE · Alternative Methods of Regression

BISGAARD and KULAHCI · Time Series Analysis and Forecasting by Example

BISWAS, DATTA, FINE, and SEGAL · Statistical Advances in the Biomedical Sciences: Clinical Trials, Epidemiology, Survival Analysis, and Bioinformatics

BLISCHKE and MURTHY (editors) · Case Studies in Reliability and Maintenance

BLISCHKE and MURTHY · Reliability: Modeling, Prediction, and Optimization

BLOOMFIELD · Fourier Analysis of Time Series: An Introduction, *Second Edition*

BOLLEN · Structural Equations with Latent Variables

BOLLEN and CURRAN · Latent Curve Models: A Structural Equation Perspective

BOROVKOV · Ergodicity and Stability of Stochastic Processes

BOSQ and BLANKE · Inference and Prediction in Large Dimensions

BOULEAU · Numerical Methods for Stochastic Processes

* BOX and TIAO · Bayesian Inference in Statistical Analysis

BOX · Improving Almost Anything, *Revised Edition*

* BOX and DRAPER · Evolutionary Operation: A Statistical Method for Process Improvement

BOX and DRAPER · Response Surfaces, Mixtures, and Ridge Analyses, *Second Edition*

BOX, HUNTER, and HUNTER · Statistics for Experimenters: Design, Innovation, and Discovery, *Second Editon*

BOX, JENKINS, and REINSEL · Time Series Analysis: Forcasting and Control, *Fourth Edition*

BOX, LUCEÑO, and PANIAGUA-QUIÑONES · Statistical Control by Monitoring and Adjustment, *Second Edition*

* BROWN and HOLLANDER · Statistics: A Biomedical Introduction

CAIROLI and DALANG · Sequential Stochastic Optimization

CASTILLO, HADI, BALAKRISHNAN, and SARABIA · Extreme Value and Related Models with Applications in Engineering and Science

CHAN · Time Series: Applications to Finance with R and S-Plus®, *Second Edition*

CHARALAMBIDES · Combinatorial Methods in Discrete Distributions

CHATTERJEE and HADI · Regression Analysis by Example, *Fourth Edition*

CHATTERJEE and HADI · Sensitivity Analysis in Linear Regression

CHERNICK · Bootstrap Methods: A Guide for Practitioners and Researchers, *Second Edition*

CHERNICK and FRIIS · Introductory Biostatistics for the Health Sciences

CHILÈS and DELFINER · Geostatistics: Modeling Spatial Uncertainty, *Second Edition*

CHOW and LIU · Design and Analysis of Clinical Trials: Concepts and Methodologies, *Second Edition*

CLARKE · Linear Models: The Theory and Application of Analysis of Variance

CLARKE and DISNEY · Probability and Random Processes: A First Course with Applications, *Second Edition*

* COCHRAN and COX · Experimental Designs, *Second Edition*

COLLINS and LANZA · Latent Class and Latent Transition Analysis: With Applications in the Social, Behavioral, and Health Sciences

CONGDON · Applied Bayesian Modelling

CONGDON · Bayesian Models for Categorical Data

CONGDON · Bayesian Statistical Modelling, *Second Edition*

CONOVER · Practical Nonparametric Statistics, *Third Edition*

COOK · Regression Graphics

COOK and WEISBERG · An Introduction to Regression Graphics

COOK and WEISBERG · Applied Regression Including Computing and Graphics

CORNELL · A Primer on Experiments with Mixtures

CORNELL · Experiments with Mixtures, Designs, Models, and the Analysis of Mixture Data, *Third Edition*

COX · A Handbook of Introductory Statistical Methods

CRESSIE · Statistics for Spatial Data, *Revised Edition*

CRESSIE and WIKLE · Statistics for Spatio-Temporal Data

CSÖRGO and HORVÁTH · Limit Theorems in Change Point Analysis

DAGPUNAR · Simulation and Monte Carlo: With Applications in Finance and MCMC

DANIEL · Applications of Statistics to Industrial Experimentation

DANIEL · Biostatistics: A Foundation for Analysis in the Health Sciences, *Eighth Edition*

* DANIEL · Fitting Equations to Data: Computer Analysis of Multifactor Data, *Second Edition*

DASU and JOHNSON · Exploratory Data Mining and Data Cleaning

DAVID and NAGARAJA · Order Statistics, *Third Edition*

* DEGROOT, FIENBERG, and KADANE · Statistics and the Law

DEL CASTILLO · Statistical Process Adjustment for Quality Control

DeMARIS · Regression with Social Data: Modeling Continuous and Limited Response Variables

DEMIDENKO · Mixed Models: Theory and Applications

DENISON, HOLMES, MALLICK and SMITH · Bayesian Methods for Nonlinear Classification and Regression

---

* Now available in a lower priced paperback edition in the Wiley Classics Library.

† Now available in a lower priced paperback edition in the Wiley–Interscience Paperback Series.

* Now available in a lower priced paperback edition in the Wiley Classics Library.

† Now available in a lower priced paperback edition in the Wiley–Interscience Paperback Series.

---

* Now available in a lower priced paperback edition in the Wiley Classics Library.

† Now available in a lower priced paperback edition in the Wiley–Interscience Paperback Series.

HUNT and KENNEDY · Financial Derivatives in Theory and Practice, *Revised Edition*
HURD and MIAMEE · Periodically Correlated Random Sequences: Spectral Theory and Practice
HUSKOVA, BERAN, and DUPAC · Collected Works of Jaroslav Hajek – with Commentary
HUZURBAZAR · Flowgraph Models for Multistate Time-to-Event Data
JACKMAN · Bayesian Analysis for the Social Sciences
† JACKSON · A User's Guide to Principle Components
JOHN · Statistical Methods in Engineering and Quality Assurance
JOHNSON · Multivariate Statistical Simulation
JOHNSON and BALAKRISHNAN · Advances in the Theory and Practice of Statistics: A Volume in Honor of Samuel Kotz
JOHNSON, KEMP, and KOTZ · Univariate Discrete Distributions, *Third Edition*
JOHNSON and KOTZ (editors) · Leading Personalities in Statistical Sciences: From the Seventeenth Century to the Present
JOHNSON, KOTZ, and BALAKRISHNAN · Continuous Univariate Distributions, Volume 1, *Second Edition*
JOHNSON, KOTZ, and BALAKRISHNAN · Continuous Univariate Distributions, Volume 2, *Second Edition*
JOHNSON, KOTZ, and BALAKRISHNAN · Discrete Multivariate Distributions
JUDGE, GRIFFITHS, HILL, LÜTKEPOHL, and LEE · The Theory and Practice of Econometrics, *Second Edition*
JUREK and MASON · Operator-Limit Distributions in Probability Theory
KADANE · Bayesian Methods and Ethics in a Clinical Trial Design
KADANE AND SCHUM · A Probabilistic Analysis of the Sacco and Vanzetti Evidence
KALBFLEISCH and PRENTICE · The Statistical Analysis of Failure Time Data, *Second Edition*
KARIYA and KURATA · Generalized Least Squares
KASS and VOS · Geometrical Foundations of Asymptotic Inference
† KAUFMAN and ROUSSEEUW · Finding Groups in Data: An Introduction to Cluster Analysis
KEDEM and FOKIANOS · Regression Models for Time Series Analysis
KENDALL, BARDEN, CARNE, and LE · Shape and Shape Theory
KHURI · Advanced Calculus with Applications in Statistics, *Second Edition*
KHURI, MATHEW, and SINHA · Statistical Tests for Mixed Linear Models
* KISH · Statistical Design for Research
KLEIBER and KOTZ · Statistical Size Distributions in Economics and Actuarial Sciences
KLEMELÄ · Smoothing of Multivariate Data: Density Estimation and Visualization
KLUGMAN, PANJER, and WILLMOT · Loss Models: From Data to Decisions, *Third Edition*
KLUGMAN, PANJER, and WILLMOT · Loss Models: Further Topics
KLUGMAN, PANJER, and WILLMOT · Solutions Manual to Accompany Loss Models: From Data to Decisions, *Third Edition*
KOSKI and NOBLE · Bayesian Networks: An Introduction
KOTZ, BALAKRISHNAN, and JOHNSON · Continuous Multivariate Distributions, Volume 1, *Second Edition*

---

* Now available in a lower priced paperback edition in the Wiley Classics Library.
† Now available in a lower priced paperback edition in the Wiley–Interscience Paperback Series.

KOTZ and JOHNSON (editors) · Encyclopedia of Statistical Sciences: Volumes 1 to 9 with Index

KOTZ and JOHNSON (editors) · Encyclopedia of Statistical Sciences: Supplement Volume

KOTZ, READ, and BANKS (editors) · Encyclopedia of Statistical Sciences: Update Volume 1

KOTZ, READ, and BANKS (editors) · Encyclopedia of Statistical Sciences: Update Volume 2

KOWALSKI and TU · Modern Applied U-Statistics

KRISHNAMOORTHY and MATHEW · Statistical Tolerance Regions: Theory, Applications, and Computation

KROESE, TAIMRE, and BOTEV · Handbook of Monte Carlo Methods

KROONENBERG · Applied Multiway Data Analysis

KULINSKAYA, MORGENTHALER, and STAUDTE · Meta Analysis: A Guide to Calibrating and Combining Statistical Evidence

KULKARNI and HARMAN · An Elementary Introduction to Statistical Learning Theory

KUROWICKA and COOKE · Uncertainty Analysis with High Dimensional Dependence Modelling

KVAM and VIDAKOVIC · Nonparametric Statistics with Applications to Science and Engineering

LACHIN · Biostatistical Methods: The Assessment of Relative Risks, *Second Edition*

LAD · Operational Subjective Statistical Methods: A Mathematical, Philosophical, and Historical Introduction

LAMPERTI · Probability: A Survey of the Mathematical Theory, *Second Edition*

LAWLESS · Statistical Models and Methods for Lifetime Data, *Second Edition*

LAWSON · Statistical Methods in Spatial Epidemiology, *Second Edition*

LE · Applied Categorical Data Analysis, *Second Edition*

LE · Applied Survival Analysis

LEE · Structural Equation Modeling: A Bayesian Approach

LEE and WANG · Statistical Methods for Survival Data Analysis, *Fourth Edition*

LePAGE and BILLARD · Exploring the Limits of Bootstrap

LESSLER and KALSBEEK · Nonsampling Errors in Surveys

LEYLAND and GOLDSTEIN (editors) · Multilevel Modelling of Health Statistics

LIAO · Statistical Group Comparison

LIN · Introductory Stochastic Analysis for Finance and Insurance

LITTLE and RUBIN · Statistical Analysis with Missing Data, *Second Edition*

LLOYD · The Statistical Analysis of Categorical Data

LOWEN and TEICH · Fractal-Based Point Processes

MAGNUS and NEUDECKER · Matrix Differential Calculus with Applications in Statistics and Econometrics, *Revised Edition*

MALLER and ZHOU · Survival Analysis with Long Term Survivors

MARCHETTE · Random Graphs for Statistical Pattern Recognition

MARDIA and JUPP · Directional Statistics

MARKOVICH · Nonparametric Analysis of Univariate Heavy-Tailed Data: Research and Practice

MARONNA, MARTIN and YOHAI · Robust Statistics: Theory and Methods

---

* Now available in a lower priced paperback edition in the Wiley Classics Library.

† Now available in a lower priced paperback edition in the Wiley–Interscience Paperback Series.

MASON, GUNST, and HESS · Statistical Design and Analysis of Experiments with Applications to Engineering and Science, *Second Edition*

McCULLOCH, SEARLE, and NEUHAUS · Generalized, Linear, and Mixed Models, *Second Edition*

McFADDEN · Management of Data in Clinical Trials, *Second Edition*

* McLACHLAN · Discriminant Analysis and Statistical Pattern Recognition

McLACHLAN, DO, and AMBROISE · Analyzing Microarray Gene Expression Data

McLACHLAN and KRISHNAN · The EM Algorithm and Extensions, *Second Edition*

McLACHLAN and PEEL · Finite Mixture Models

McNEIL · Epidemiological Research Methods

MEEKER and ESCOBAR · Statistical Methods for Reliability Data

MEERSCHAERT and SCHEFFLER · Limit Distributions for Sums of Independent Random Vectors: Heavy Tails in Theory and Practice

MENGERSEN, ROBERT, and TITTERINGTON · Mixtures: Estimation and Applications

MICKEY, DUNN, and CLARK · Applied Statistics: Analysis of Variance and Regression, *Third Edition*

* MILLER · Survival Analysis, *Second Edition*

MONTGOMERY, JENNINGS, and KULAHCI · Introduction to Time Series Analysis and Forecasting

MONTGOMERY, PECK, and VINING · Introduction to Linear Regression Analysis, *Fifth Edition*

MORGENTHALER and TUKEY · Configural Polysampling: A Route to Practical Robustness

MUIRHEAD · Aspects of Multivariate Statistical Theory

MULLER and STOYAN · Comparison Methods for Stochastic Models and Risks

MURTHY, XIE, and JIANG · Weibull Models

MYERS, MONTGOMERY, and ANDERSON-COOK · Response Surface Methodology: Process and Product Optimization Using Designed Experiments, *Third Edition*

MYERS, MONTGOMERY, VINING, and ROBINSON · Generalized Linear Models. With Applications in Engineering and the Sciences, *Second Edition*

NATVIG · Multistate Systems Reliability Theory With Applications

† NELSON · Accelerated Testing, Statistical Models, Test Plans, and Data Analyses

† NELSON · Applied Life Data Analysis

NEWMAN · Biostatistical Methods in Epidemiology

NG, TAIN, and TANG · Dirichlet Theory: Theory, Methods and Applications

OKABE, BOOTS, SUGIHARA, and CHIU · Spatial Tesselations: Concepts and Applications of Voronoi Diagrams, *Second Edition*

OLIVER and SMITH · Influence Diagrams, Belief Nets and Decision Analysis

PALTA · Quantitative Methods in Population Health: Extensions of Ordinary Regressions

PANJER · Operational Risk: Modeling and Analytics

PANKRATZ · Forecasting with Dynamic Regression Models

PANKRATZ · Forecasting with Univariate Box-Jenkins Models: Concepts and Cases

PARDOUX · Markov Processes and Applications: Algorithms, Networks, Genome and Finance

PARMIGIANI and INOUE · Decision Theory: Principles and Approaches

---

  * PARZEN · Modern Probability Theory and Its Applications

     PEÑA, TIAO, and TSAY · A Course in Time Series Analysis

     PESARIN and SALMASO · Permutation Tests for Complex Data: Applications and Software

     PIANTADOSI · Clinical Trials: A Methodologic Perspective, *Second Edition*

     POURAHMADI · Foundations of Time Series Analysis and Prediction Theory

     POWELL · Approximate Dynamic Programming: Solving the Curses of Dimensionality, *Second Edition*

     POWELL and RYZHOV · Optimal Learning

     PRESS · Subjective and Objective Bayesian Statistics, *Second Edition*

     PRESS and TANUR · The Subjectivity of Scientists and the Bayesian Approach

     PURI, VILAPLANA, and WERTZ · New Perspectives in Theoretical and Applied Statistics

  † PUTERMAN · Markov Decision Processes: Discrete Stochastic Dynamic Programming

     QIU · Image Processing and Jump Regression Analysis

  * RAO · Linear Statistical Inference and Its Applications, *Second Edition*

     RAO · Statistical Inference for Fractional Diffusion Processes

     RAUSAND and HØYLAND · System Reliability Theory: Models, Statistical Methods, and Applications, *Second Edition*

     RAYNER, THAS, and BEST · Smooth Tests of Goodnes of Fit: Using R, *Second Edition*

     RENCHER and SCHAALJE · Linear Models in Statistics, *Second Edition*

     RENCHER and CHRISTENSEN · Methods of Multivariate Analysis, *Third Edition*

     RENCHER · Multivariate Statistical Inference with Applications

     RIGDON and BASU · Statistical Methods for the Reliability of Repairable Systems

  * RIPLEY · Spatial Statistics

  * RIPLEY · Stochastic Simulation

     ROHATGI and SALEH · An Introduction to Probability and Statistics, *Second Edition*

     ROLSKI, SCHMIDLI, SCHMIDT, and TEUGELS · Stochastic Processes for Insurance and Finance

     ROSENBERGER and LACHIN · Randomization in Clinical Trials: Theory and Practice

     ROSSI, ALLENBY, and McCULLOCH · Bayesian Statistics and Marketing

  † ROUSSEEUW and LEROY · Robust Regression and Outlier Detection

     ROYSTON and SAUERBREI · Multivariate Model Building: A Pragmatic Approach to Regression Analysis Based on Fractional Polynomials for Modeling Continuous Variables

  * RUBIN · Multiple Imputation for Nonresponse in Surveys

     RUBINSTEIN and KROESE · Simulation and the Monte Carlo Method, *Second Edition*

     RUBINSTEIN and MELAMED · Modern Simulation and Modeling

     RYAN · Modern Engineering Statistics

     RYAN · Modern Experimental Design

     RYAN · Modern Regression Methods, *Second Edition*

     RYAN · Statistical Methods for Quality Improvement, *Third Edition*

     SALEH · Theory of Preliminary Test and Stein-Type Estimation with Applications

     SALTELLI, CHAN, and SCOTT (editors) · Sensitivity Analysis

     SCHERER · Batch Effects and Noise in Microarray Experiments: Sources and Solutions

  * SCHEFFE · The Analysis of Variance

*Now available in a lower priced paperback edition in the Wiley Classics Library.

† Now available in a lower priced paperback edition in the Wiley–Interscience Paperback Series.

---

* Now available in a lower priced paperback edition in the Wiley Classics Library.
† Now available in a lower priced paperback edition in the Wiley–Interscience Paperback Series.